OIL
AND POLITICS IN MODERN BRAZIL
PETER SEABORN SMITH

MACMILLAN OF CANADA/MACLEAN-HUNTER PRESS

©1976, The Macmillan Company of Canada Limited

All rights reserved—no part of this book may be reproduced in any form without permission in writing from the publisher, except by a reviewer who wishes to quote brief passages in connection with a review written for inclusion in a magazine or newspaper.

ISBN 0-7705-1286-0

Printed in Canada for
The Macmillan Company of Canada Limited
70 Bond Street
Toronto M5B 1X3

CONTENTS

Preface	v
Maps: States and Territories of Modern Brazil	vii
Principal Sedimentary Basins of Brazil	viii
Recôncavo Basin, with principal oilfields to 1960	ix
Terms and Abbreviations	xi
Introduction	1
1 Era of Private Enterprise, 1864–1930	7
2 Government Becomes Concerned, 1930–43	20
3 Liberalism Thwarted, 1943–50	46
4 Economic Nationalism Triumphant, 1950–54	75
5 Years of Success, 1954–60	102
6 Political Turmoil and the Stagnation of Petrobrás, 1960–64	133
7 Reassertion of Aims, 1964–70	165
Postscript: Self-sufficiency by 1979?	188
Appendix I Decree-Law 538, July 7, 1938	190
Appendix II Draft Proposal of the Petroleum Statute (1947): Excerpts	196
Appendix III Law 2004, October 3, 1953	207
Notes	221
Bibliography	263
Index	277

PREFACE

In the aftermath of the 1930 revolution, Brazil's new Minister of Agriculture observed that Brazil constituted half the territory of South America. This unequivocal fact was, however, scandalous to the Minister; Brazil had not found any oil within its national boundaries. "Almost all its neighbours have found that they possess oil," emphasized this spokesman of the new order, "some in formidable quantities. Does only Brazil not have it? Finding out is a question of national honour."[1]

For the past forty years the Brazilian search for oil has been technically as well as metaphorically a national enterprise, its story unique outside the Communist Bloc. In every other oil-producing country in this region, private companies — usually of foreign origin and more often than not huge vertically integrated concerns operating on a global scale — have been the instigators of oil exploration.

State oil companies in the non-Communist world, with the exception of Brazil, have been afterthoughts, coming into being only after a national oil industry — or simply oil production — had been firmly established by private-company efforts. Such state companies operate in a variety of ways: 1) as controlling agencies, supervising activities by firms under contract; 2) in company with private corporations, national or foreign; and 3) monopolizing the industry, usually after nationalization.

Brazil is an exception to the norm. Exploration did not begin under the aegis of foreign private companies; indeed, one could argue convincingly that *Brazilian* private companies did not even search for oil, since their efforts were so ill planned, poorly financed, and unproductive. The search for oil, in fact, resulted more from government stimulus than any other impulse. Moreover, the Brazilian oil industry was nationalized before oil had been discovered. Finally, the industry was put in the hands of a virtual state-monopoly company, Petróleo Brasileiro Sociedade Anônima, before the industry was of a significant size. This company has discovered and exploited most of the oilfields now producing in Brazil, and has spent hundreds of millions of dollars and cruzeiros searching the national territory for more. The Brazilian oil industry thus owes its existence to the state; it is an anomaly in the western world.[2] The study that follows explores the development of this unique situation and discusses the results — both good and bad — of it, from the late nineteenth century to the current decade.

I have many debts to acknowledge, but limitations of space and scholarly absence of mind preclude my naming most of the people and institutions in Canada, the United States, and Brazil who helped me. They know who they are and they know I am grateful. I should, however, like to single out Dr. Glycon de Paiva Teixeira and Dr. Irnack Carvalho do Amaral, both of Rio de Janeiro, for special thanks. The Henry L. and Grace Doherty Charitable Foundation, Inc., and the Canada Council honoured me with grants which permitted extended stays in Brazil in order to do research for this and related studies. I am deeply grateful to both institutions. The book has been published with the help of a grant from the Social Science Research Council of Canada, using funds provided by the Canada Council. Finally, my wife, Anne, has given me unflagging support throughout the preparation of this manuscript. My deepest gratitude to her.

P.S.S.
Rio de Janeiro, 1974

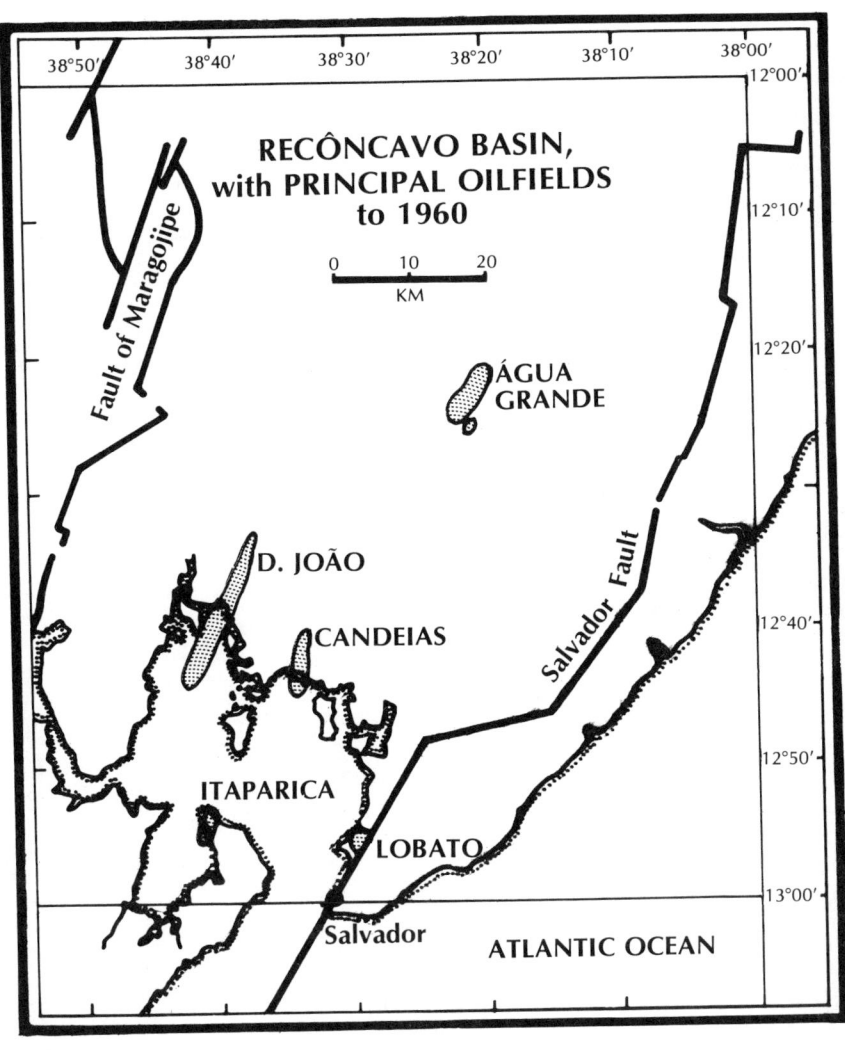

TERMS AND ABBREVIATIONS USED IN THE TEXT

Assessoria Econômica Assessoria Econômica da Presidência da República (Economic Advisory Board of the Presidency of the Republic). A "brains trust" that Getúlio Vargas set up after his election in 1951.
BNDE Banco Nacional de Desenvolvimento Econômico (National Bank of Economic Development)
b.p.d. Barrels per day
CGT Commando Geral dos Trabalhadores (General Labour Command)
CNP Conselho Nacional do Petróleo (National Petroleum Council)
conto Conto de reis=1000 milreis=1,000,000 reis. Monetary units in Brazil until 1940, at which time the conto was worth U.S. $50.
Cr$, cruzeiro Monetary unit of Brazil, 1940–67. In 1940, 1 cruzeiro=1 milreis, valued at 20 to the U.S. dollar; in 1967, it was valued at 2700 to the U.S. dollar.
DEPEX Exploration Department of Petrobrás
DEPRO Production Department of Petrobrás
DNPM Departamento Nacional da Produção Mineral (National Department of Mineral Production). Its forerunner (July 1933–March 1934) was the Departamento Geral da Produção Mineral (General Department of Mineral Production).
entreguismo Pejorative term used to describe the policy of those Brazilians ("entreguistas") who are accused of advocating excessive foreign influence in Brazil
Estado Nôvo "New State" founded in November 1937 by Getúlio Vargas, patterned after the Portuguese regime instituted in 1933 by Antônio Salazar

FPN   Frente Parlamentar Nacionalista (Nationalist Parliamentary front)
FRONAPE   Frota Nacional de Petroleiros (National Tanker Fleet)
getulista   A follower of Getúlio Vargas
IMF   International Monetary Fund
interventor   Official replacing a state governor during the Provisional Government of Getúlio Vargas, 1930–34
Law 2004   The law creating Petrobrás; see Appendix III.
Link Report   A 1960 assessment of Brazil's oil potential; attributed by radical nationalists solely to Walter K. Link, it was actually a collaborative product of Petrobrás's DEPEX.
Linkismo   Pejorative term used to describe the policy of those Brazilians (*"Linkistas"*), who were accused of trying to sabotage Petrobrás
NCr$, New cruzeiro   Monetary unit of Brazil, 1967–70, arrived at by dividing cruzeiros by 1000. (The monetary unit is now known simply as cruzeiro.)
"O petróleo é nosso"   "The oil is ours"
PCB   Partido Comunista Brasileiro (Brazilian Communist Party)
Petroleum Centre   Centro de Estudos e Defesa do Petróleo e da Economia Nacional (Study and Defence Centre of Petroleum and the National Economy)
Petroleum Statute   Estatuto do Petróleo; a liberal oil policy proposed in 1947 and defeated by the "O petróleo é nosso" campaign
Petrobrás   Petróleo Brasileiro Sociedade Anônima
PR   Partido Republicano (Republican Party)
PSB   Partido Socialista Brasileiro (Brazilian Socialist Party)
PSD   Partido Social Democratica (Social Democratic Party)
PSP   Partido Social Progressista (Social Progressive Party)
PTB   Partido Trabalhista Brasileiro (Brazilian Labour Party)
Roboré Accord   A 1958 agreement between Brazil and Bolivia setting aside a zone in Bolivia where oil could be exploited for Brazilian consumption
SALTE   A federal development plan proposed in 1948. Its name derives from the first letters of the Portuguese words for health (*saúde*), food (*alimentação*), transportation (*transporte*), and energy (*energia*).

SFPM   Servico de Fomento da Produção Mineral (Service for the Stimulation of Mineral Production)
SGM   Servico Geológico e Mineralógico (Geological and Mineralogical Service)
tenente   "lieutenant"; any of the handful of junior officers who participated in the 1922 Fort Copacabana revolt. *Tenentismo* described their apparent creed, which was basically corporatism.
"trusts"   Pejorative term for the great international petroleum companies
UDN   União Democrática Nacional (National Democratic Union)
UNE   União Nacional dos Estudantes (National Student Union)

In accordance with common practice, I have used interchangeably the words "petroleum" and "oil". Although in 1940 Brazilian Portuguese underwent an orthographic change, old spellings sometimes appear, notably with names. In the text I have consistently used the most common spelling, usually the modern.

INTRODUCTION

This book describes the emergence and evolution of the Brazilian oil industry, from its tentative beginnings late in the nineteenth century to maturity in 1970. The book is a study of how the scarcity of capital and expertise characteristic of an underdeveloped country and the nationalism equally characteristic of such countries in the mid-twentieth century interacted with the facts of geology to produce a uniquely Brazilian result. To elaborate on this theme, scarcity of capital meant that Brazil had but two options for creating a national oil industry: private foreign capital (from the international oil companies) or state funds; there was not enough private Brazilian capital to tackle the enterprise. Scarcity of expertise meant that little or nothing was known about Brazil's petroleum geology until very recent times, and that meanwhile, faith that Brazil was rich in oil prevailed in the public's mind. Nationalism and oil became inextricably linked during the 1930s, as Brazil industrialized and sought energy sources; primarily because of the lack of information about the nation's petroleum geology, oil nationalism took on a mythic quality: Brazilians believed that their subsoil was rich in oil and that the international oil companies would stop at nothing to get it. The result of all these factors was state control of oil before it had been discovered in Brazil, and a state-monopoly company before an oil industry worthy of the name had been created. This book describes how and why this unusual process occurred.

Much of the study that follows deals with public opinion, about which a point must be made: it is not so much the truth that matters as what people think the truth is. This book shows that the Brazilian oil industry assumed its particular form largely as a result of public opinion, which was based on two beliefs—myths—that had little basis in fact: 1) that Brazil was rich in oil, and 2) that the international oil companies were eager to exploit Brazil's reserves. These myths emerged initially in a vacuum—there was virtually no genuine information about Brazil's petroleum geology available until at least the Second World War—and they have persisted because information to the contrary was too little and too late in coming to stop the myths being established as "truth".

Before examining Brazil's petroleum geology, a brief layman's introduction to the subject in general is perhaps in order. Petroleum was probably formed from the remains of tiny aquatic plants and animals that lived many millions of years ago, when the seas were much more extensive than they are now. The remains, mixed with the mud and sand of the seabed, were covered and recovered with new layers of the same "marine sediment"; over the centuries, pressure from succeeding layers turned the older sediments into rock. From time to time the earth's crust buckled, distorting the layers of sediment; the heat and pressure turned the marine sediments into oil. This petroleum can be exploited economically only when it has been concentrated into pools or fields of substantial capacity (usually a million barrels — of thirty-five imperial or forty-two U.S. gallons — as a minimum); such concentrations occur when oil moves through porous rock (limestone, dolomite, sandstone) into a trap, where it is imprisoned by layers of impermeable rock (clay or shale) above and below it. The most common means by which a trap is created is folding, which produces a dome or anticline of hard rock that forces the layers of sediments above it into a similar humped configuration. Sometimes the earth's crust will crack through several layers of rock, and the crust on one side of the crack or fault will be thrust against the other; a layer of oil-bearing sedimentary rock may then be thrust against impermeable rock, again creating a trap. There are other geological movements that can trap oil, but the most important is the aforementioned dome, because all the petroleum that migrates from more distant holdings is contained by the impermeable cap rock. The thicker the layer of porous rock, the greater the yield of oil; the more impermeable the layers above and below the oil-bearing sedimentary rock, the greater the percentage of the oil trapped in the reservoir.

To turn to Brazil, roughly 40 per cent of its territory comprises sedimentary basins of various types and thicknesses; this fact alone has been used to support assertions that Brazil must hold vast deposits of petroleum. The basins may be grouped into four "oil provinces": Amazon (Upper, Middle, and Lower), Middle North (Maranhão, Barreirinhas), Atlantic (Northeast Coastal, Recôncavo, Tucano, Espírito Santo, and Campos), and Paraná. The Amazon, Middle North, and Paraná have extensive areas, are located near mountains (hence they should exhibit folding, influenced by the mountain-building), and contain structures associated elsewhere in the world with petroleum produced by "tectonic" (mountain-building) distortions.

But all four provinces lack some important characteristics normally encountered in oilfields of high productivity. In the first place, none of the basins is of more than medium depth, and the sedimentary layers in them are not particularly thick. Then, too, Brazilian geological history lacks folding of the sort associated with mountain-building within thick sedimentary beds, through horizontal petroleum zones, such as occurs in most of the important oil-producing regions of the world. The basins, moreover, do not show an unvaried alternation of porous and impermeable rocks; formations are irregular and difficult to perceive owing to the weakness of the rock structures. (This means that not only will oil migrate with great difficulty, but that traps will be weak, retaining relatively little oil.) Furthermore, basaltic lava intrudes into the sediments in the basins of all the provinces except the Atlantic, pressing the sediments together and interrupting the flow of petroleum through the structures. Such igneous rocks also cover great areas of the three provinces — particularly Paraná — creating a barrier to exploration virtually insurmountable with the technology of the era covered in this book. To sum up, Brazilian geology is not conducive to the accumulation of substantial fields of oil, and the discovery and extraction of the oil that is there would be relatively difficult and expensive.[1]

The reader should be aware, however, that he or she now

knows more about Brazil's petroleum geology than virtually all Brazilians knew before 1960, with the exception of people directly involved with the industry and those who read the few articles that had appeared in technical journals and the scant information divulged in official reports up to that time. Neither the state-control agency, created in 1938, nor the state-monopoly company that began operations in 1954 divulged much technical information about the petroleum geology their explorations were uncovering, preferring to relate statistics about drilling, production, and other phases of the industry. Given the relatively limited program of exploration that was carried out prior to 1954, owing chiefly to small budgets, there was all too little information to divulge. It seems to have become an article of faith among Brazilians by the 1930s that their nation's vast sedimentary basins held equally vast quantities of petroleum, a belief that was not openly challenged by published official evidence until 1960. And, as this book relates, faith prevailed for some years thereafter.

Finally, a word about sources. This study relies heavily on newspaper opinion for the period after the Second World War, using newspapers almost exclusively from Rio de Janeiro, the official capital of Brazil until 1960 and the unofficial capital for some years thereafter. The many Rio dailies differ in quality as well as opinion, and one may fairly sample the spectrum of Brazilian public opinion on any issue from them. Since this book is concerned with nationalistic opinion and its influence upon events, the radical-nationalist daily, *Última Hora*, was consulted heavily for the years during which radical nationalism was an effective force. On specific issues, newspapers representing different opinions and of different quality were consulted; for example, liberal (representing the traditional, export-oriented business sector and people who supported the widespread participation of foreign capital in the Brazilian economy): *O Globo* and *O Jornal* (after 1950). The *Jornal do Brasil* was and is a newspaper of high quality, comparable in reporting and readership to the *New York Times*; it was useful for, among other things, careful assessment of contentious issues, particularly between 1961 and 1964 when opinion polarized and rationale became increasingly difficult for the outsider to understand. Such sampling was supplemented, as the bibliography indicates, by other newspapers and journals, and by the excellent clippings file found in the IBGE.

This book, written by a historian, is a work of history, a monograph. As such, it describes the events pertaining to a particular subject and analyses the events in the course of the narrative; the analysis is in accordance with the historical evidence discovered and, inevitably, with the author's own interpretation of that evidence. The book is intended for a wider audience than historians, however, and should be of particular interest to students of development. But it does not construct models against which to test the evidence; such is more properly the work of political scientists, economists, and sociologists, who should be able to use this "case" in their explorations of the controversial subject of development. Moreover, the author has deliberately avoided, for example, a discussion of the merits and drawbacks of foreign investment in a developing country; that would be the subject of a different and much longer book. What is contained in this monograph is an example of one nation's reaction to foreign investment in an important sector of the economy.

1
ERA OF PRIVATE ENTERPRISE, 1864–1930

One of the few areas in Brazil where oil seepages are found is near the city of Salvador, Bahia. Natives in that region used oil- and resin-covered arrows to set fire to the blockhouses of the Portuguese invader as well as the huts of other tribes in warfare. In the mid-nineteenth century British engineers building the Bahia–São Francisco Railway were excited by marine fossils they encountered in cuts along the right-of-way, and they spread word that the region looked promising for petroleum.

In 1822 Brazil won her independence from Portugal and became an empire. Mining legislation was a continuation of colonial policy: subsoil wealth was considered the property of the Imperial Crown (in place of the Portuguese Crown in colonial times). Thus, in granting exploration rights the imperial government would stress to the concessionaire that he would have to seek exploitation rights under a separate concession. As evidence of some interest in prospecting, the imperial government established a mining school in Minas Gerais, which opened in 1876. Brazil was included in the world-wide search for combustible hydrocarbons — chiefly coal — in the nineteenth century, and many foreigners made geological studies of various Brazilian regions.

It was not until late in the nineteenth century that anyone began to search specifically for oil. Two Englishmen, Thomas Denny Sargent and Edward Pellew Wilson, were the first; Sargent received a ninety-year exploration concession around Ilheus and Camaú, Bahia, and Wilson was granted a similar concession along the Maraú River, in the same state. Nothing more is known of Sargent, but Wilson achieved brief notoriety.

First, landowners within his concession disputed the government's right to cede from their holdings; Wilson's concession was upheld, however, and he was able to acquire others in the area. Then he formed a company and built a distillation plant in 1884, hoping to glean oil from the bituminous schists around Maraú, but John Cameron Grant, manager and part owner of the plant, proved unable to get along with his employees; indeed, he shot and killed one. Although exonerated in the subsequent trial, Grant left Brazil and the plant closed in 1893, after a reputed expenditure of half a million pounds sterling on the enterprise.

At the same time other individuals were prospecting for oil in Brazil, mainly in the south. Most concentrated their search in São Paulo state, but two intrepid pioneers did venture into Maranhão, and one went south into Santa Catarina. This last explorer apparently marched over a wide area with some treatises on geology under his arm, searching for traces of petroleum and talking continually of anticlines and diabase; his frustration eventually drove him mad. None of these men drilled, but they searched for seepages and other surface traces.[1]

In 1889 the Brazilian Empire became a republic. The new constitution, promulgated in 1891, reversed a long-standing trend toward centralization of government, creating a rather loose federal system. This clearly reflected the interests of the great landholders, "planters", who controlled Brazil's completely agricultural economy and who wanted as little governmental interference in their affairs as possible. Further proof of their influence was found in the clauses regulating mining. Landholders had resented the imperial government's granting of exploration rights for subsoil deposits, even under privately owned land; under the new constitution those who owned the soil owned the subsoil and its wealth. The constitution also turned over mine ownership and the regulation of mining to state governments.[2] At this time planters — who were not interested in mining — controlled state governments; thus, the immediate effect of this legislative change was a sharp drop in minerals exploration and development, and the industry did not recover until after the turn of the century. For example, an English company, which had brought a drilling rig to the

Alagoas coastline just before promulgation of the constitution, immediately thereafter abandoned the site and the country. Only two states, São Paulo and Minas Gerais, both of which had a historic interest in mining, created geological services, but neither was able to stir interest in exploration in their own territories.

A noteworthy exception to the norm occurred in São Paulo, however, where in 1892 Eugênio Ferreira Camargo, himself a *Paulista* (native of São Paulo), bought the Bofete *fazenda* (plantation) from a company that had gone bankrupt trying to drill for oil on it. Camargo expended his fortune drilling on his own land, reaching a depth of between 400 and 486 metres by 1896. His neighbours believed him to be in league with the devil, since he was bringing smoke out of the bowels of the earth. The result of his efforts was not oil, however, but sulphurous water, which continues to flow today. Brazilian commentators hold Camargo to be the national oil pioneer, the first native to expend his own money in the search for oil in Brazil.[3]

By the turn of the twentieth century it had become clear to a few Brazilians that the federal government would somehow have to prod the flagging mining industry. In 1901 Alfredo Maia, Transport Minister in the Campos Sales federal administration, polled state governments on reasons for the lack of mining activity; the caprice or inertia of the landowner and confusion over boundaries and rights were the factors most often cited.[4]

Yet the constitution precluded any direct federal activity in mining, unless provided for by law in order to regulate the industry. One legislator in particular, João Pandiá Calógeras, began at this time to promote federal intervention in mining. Although he was later to write works on the subject — vast catalogues of national subsoil wealth — and be elevated twice to the post of Minister of Agriculture (which had jurisdiction over mining), his efforts to formulate a mining code did not bear fruit for decades. For a number of reasons, not the least of which was planter resistance to any increase in federal power, politicians persistently blocked federal intrusion into mining.[5]

One thing the federal government could do was conduct surveys, and in 1903 the Rodrigues Alves administration hired an American geologist, Israel Charles White, to survey combustible minerals in the south of Brazil. A former director of the Virginia Geological Service, White investigated the coal possibilities of Rio Grande do Sul and Santa Catarina, under which the Paraná sedimentary basin lies (which unfortunately is overlain by igneous rocks). In his report, published in 1908, White devoted only two pages to the oil potential of the region and was frankly pessimistic. He also claimed, however, that another geologist had tried to pass off a sample of refined oil as crude extracted from southern Bahia, the region which had first attracted international attention.[6] Whether White was correct or not in his analysis of the sample, he would not be the last to dismiss claims about oil in Bahia in that way.

In another effort to promote surveying, the Affonso Pena administration created the Geological and Mineralogical Service of Brazil (SGM), under the Ministry of Agriculture, and brought in the American geologist Orville Derby to head it. His task proved far more difficult than expected, and he became so frustrated and depressed by the restrictions local planters and officials continually placed in his path that he committed suicide after a few years as director.[7]

The First World War awakened world interest in oil. The mechanization of armies, the creation of air forces, and the replacement of coal with petroleum by the world's navies presented new problems to military strategists and sharply increased world demand for oil. Similarly, the automobile had by this time come into its own, firmly establishing the internal-combustion engine as a compact, economical source of energy. A few automobiles had been seen in Brazil by the time war broke out, but the petroleum derivatives they consumed had to be imported.

This internationally awakened interest in oil was somewhat slow to reach Brazil. For instance, although the SGM's new director, Brazilian geologist Euzébio Paulo de Oliveira, encouraged the study of oil in 1915, his chief purpose was apparently to refute Israel White's pessimistic assessment of the Brazilian south, principally because White was a foreigner.[8] Oliveira, a staunch nativist, would later strongly influence Brazil's emerging oil policy. At the same time, the SGM came under attack. In a confidential report submitted in 1918, Calógeras characterized the agency as a feather bed, with too many incompetent employees and with little internal cohesion or direction.⁹

In 1917 the SGM got a new director, Brazilian geologist Gonzago de Campos, and began to take an interest in oil exploration. Campos immediately created a commission to study the coal and oil possibilities of the Amazon Valley. He also appealed to the Ministry of Agriculture for funds to begin drilling for oil. The Ministry responded by establishing a department for oil exploration within the SGM.

This new department was created to remedy the deficiencies of private enterprise. The incident that prompted its formation was the imminent collapse of the Empresa Paulista de Petróleos, formed in 1917 to drill in São Paulo state. Unable even to begin operations because it had no money, the company appealed to the Ministry of Agriculture. The Ministry bought a drilling rig, rented it to the company, and provided supervisory technicians (as did the Geological Service of São Paulo). The company was then able to amass sufficient capital (the equivalent of $70,000) to drill to a depth of only 300 metres before it went bankrupt.¹⁰

Despite the fact that this venture failed, it brought the Brazilian government into petroleum exploration to stay. It also set the pattern for government activity for the next decade: provision of equipment and technical expertise to private companies after they had acquired capital and selected a drilling site. All too often, unfortunately, companies embarked on exploration and selected drilling sites with little or no knowledge of petroleum geology or technology.

While the Empresa Paulista de Petróleos was collapsing and the Amazon commission was exploring the jungle, a young man claimed to have discovered oil in the Riacho Doce region of Alagoas. Although it soon became apparent that he was quite mad, his claim struck a responsive chord in SGM Director Gonzago de Campos, who persuaded the federal government to appropriate funds for exploration there.¹¹

In 1919 the SGM decided to try drilling a well on its own, in Paraná. The attempt was hardly noteworthy: SGM drillers abandoned the site after eleven months, having reached a depth of only eighty-four metres; so dismayed was Gonzago de Campos at the amateurish effort that characterized oil exploration in

Brazil that he appealed to the federal government for permission to hire foreign drilling companies. Not only was permission refused, but Campos was dismissed as SGM director and replaced by Euzébio Paulo de Oliveira. The reappointment of Oliveira underscored the persistent attitude of nativism on the part of the federal government: minerals in general, and oil in particular, were to be sought exclusively by Brazilians or not at all. Postwar world-wide mistrust of giant foreign monopolies probably reinforced this attitude, combined with the chauvinistic 1917 Mexican Constitution and growing Brazilian national pride as the centenary of independence approached in 1922.

Oliveira himself has been described as "deeply patriotic, the epitome of chauvinism" and "totalitarian". In his *Bulletin Number 1* of 1919 Oliveira set forth two goals for Brazilian oil exploration: 1) the location of drilling sites by native geologists, and 2) drilling by Brazilians. For almost twenty years he pursued these goals with rare single-mindedness, tempered by an awareness of political reality.[12] In 1923, for example, he proposed a law prohibiting concessions to foreigners within sixty kilometres of the coast and national boundaries. The proposal further required that at least 25 per cent of stock in oil companies exploring in Brazil be held by Brazilians, and that at least two concession areas in each oilfield be reserved for national exploitation. Within the oil companies working in the national territory, a specified percentage of native personnel was to be employed in both administrative and technical work. Exploration concessions were to be from two to four years in duration, with no time limit on exploitation concessions. The various companies were to build schools and hospitals for workers.[13] While no action was taken on Oliveira's proposal, his intentions are clear. Banning foreigners from exploration along national limits might make the states affected consider exploration in those areas, hopefully by the SGM, and the nationalistic strictures placed upon foreign companies exploring anywhere else in Brazil should dissuade them from entering, leaving the activity to Brazilians.

Oliveira also managed to increase the meagre appropriation the SGM received annually from the federal government. When he assumed office in 1919, it amounted to slightly less than the equivalent of $100,000; by the mid-1930s it was five times that amount. Still, it was pathetically small. During the 1920s the SGM drilling rigs often could not be repaired, and only rarely were they replaced or was additional equipment purchased; the 1924 allocation, for example, "scarcely paid the cost of one drilling", according to a contemporary observer.[14]

The appointment of so single-minded an individual as Oliveira also meant that government oil exploration was to be directed by geologists, not engineers. Since geologists are relatively ignorant of drilling techniques, Oliveira's appointment had the effect of retarding the development of drilling. It was further retarded because Oliveira did not allow his technicians to travel and observe new practices, nor did his chauvinism allow him to permit foreigners to come into contact with his organization. The SGM thus could not hire foreign personnel, even as a temporary measure, until enough Brazilians could be trained. It must be pointed out, however, that Oliveira did create the organization from which eventually came a body of technicians capable of fulfilling government drilling contracts.[15]

As if the lack of money and qualified personnel were not a sufficient restraint on the SGM, there was another factor: restrictive legislation. The chief stumbling-block was, of course, the 1891 Constitution, which put subsoil wealth into the hands of landowners. Although such a legal regime operated to the benefit of the oil industry in such countries as the United States, it had the opposite effect in Brazil, principally because of the historic disinclination of the Brazilian landowner to invest in such high-risk ventures as oil exploration. Most were content to reap their profits from coffee or sugar and deposit them in a safe place. Moreover, Brazilian businessmen were hardly economic nationalists, being concerned chiefly with exporting raw materials and importing finished goods; they saw no need to create a national oil industry. What point was there in worrying about a national supply of energy, since there was little industry to consume it and adequate supplies of coal and hydro-electric power existed to supply current needs? Brazil in the 1920s also made foreign oil explorers unwelcome; they were not prevented from coming, but received neither support nor co-operation in their efforts. Further, the Ministry of Agriculture prohibited foreign prospectors not only from discussing their findings while in Brazil, but also from mentioning the foreign oil industry to anyone connected with the Ministry.[16]

Another group that might have been expected to have shown some interest in Brazil's sedimentary basins comprised the international oil companies. That they did not was probably due to several factors: 1) despite the fact that roughly 40 per cent of Brazilian territory contains sedimentary basins of various types and thicknesses, few have undergone substantial folding and faulting of the sort associated with mountain-building, which traps oil in great pools; 2) the sheer size of the terrain to be surveyed; and 3) the greater attraction of such countries as Venezuela, in which the national government owned subsoil deposits and was eager to have them exploited. Only the state governments could have enticed the great international oil companies to Brazil, since the 1891 Constitution also gave them title to vacant lands within their territories. Planter domination of such governments, however, precluded such a possibility.

In such a climate of, at best, indifference to oil exploration and, at worst, hostility, probably the only entity that could have usefully carried out oil prospecting was the government, through the SGM. But the SGM could explore only on federal land, principally the territories, remote and difficult of access.

Federal authorities tried to encourage exploration by private parties, but with scant success. In 1915, for example, João Pandiá Calógeras, who had tried repeatedly to prod Brazilians' interest in mining, drafted a bill that permitted a third party, known as an *inventor*, to explore on private land so long as he agreed to share any eventual profit with the owner. While Calógeras's bill failed, Ildefonso Simões Lopes, Minister of Agriculture from 1919 to 1922 and a staunch nationalist, managed to have a similar measure enacted in 1921. His law also stipulated that privately owned oil wells could be expropriated by either state or federal governments should "public necessity" so dictate. The new arrangement had little effect on exploration: few prospective *inventores* had sufficient capital and interest to risk a venture (federal-government entities were ineligible), and lack of capital hampered the SGM.[17]

Thus, Brazil's search for oil into the 1920s suffered from three deficiencies: capital, trained personnel, and legislative inducements. The effort had been sporadic and fitful, and it is doubtful that the nation was really interested in finding and developing a national source of oil. President Artur Bernardes (1922–26), for example, was concerned more with excluding foreigners from the activity than with allocating sufficient funds to permit more than token exploration. Indeed, Bernardes followed the policy stated by Simões Lopes when the latter was Minister of Agriculture: the federal government must explore in order to restrict foreign incursions. Clearly, oil exploration was not particularly important to Brazil at this time.

Apart from the few efforts of the SGM, therefore, exploration remained the whim of the amateur. For example, in 1925 Pinto Martins of Ceará, the first aviator to link the three Americas by air, decided to have a fling at oil exploration in Alagoas. He was soon found dead, reportedly by suicide (another suicide provoked by the search for oil in Brazil!), in a hotel room in Rio de Janeiro.[18] Other individuals, no more qualified, were exploring in the same general area and with as little success, provoking ridicule of Brazil's exploration efforts such as that provided in the *Oil and Gas Journal* early in 1925. The writer cited inadequate drilling, both in number of holes and in depth, antiquated equipment, and the lack of experienced drillers and scientific judgement in selecting drilling sites. He doubted that oil would ever be found in Brazil.[19]

Within a year the São Paulo government, apparently prompted by an emerging world glut of coffee, its chief industry, decided to commission a survey of the state's oil potential. The state hired an American geologist, Dr. Chester W. Washburne, to conduct the survey; the federal SGM also contributed a geologist to the task. After working in the field for three years, Washburne submitted his report to the state Secretary of Agriculture. The geologist was optimistic on theoretical grounds: based on the shallow drilling already done, he inferred the presence of oil-bearing rocks of Devonian origin at deeper levels than had yet been explored — despite the presence of igneous rock, overlying and intruding into the sedimentary formation. He called for exploration by deep-drilling rigs, which should encounter this formation and serve to attract oil companies. Washburne asserted that the central and southern parts of the Paraná Basin of São Paulo offered the best prospects for deep drilling (to at least 4,000 feet), but warned that "many years of geologic study and of drilling" would probably be necessary before any profitable oilfields could be developed.[20]

São Paulo was the only state in Brazil to attempt a survey of its oil potential; Washburne's conclusions were speculative at best, however, because he had to rely on the results from drilling up to that time and did little or no drilling himself. The story is told that Julio Prestes, Governor of São Paulo, informed Washburne that he had 3,000 contos in funds for his survey. Upon leaving Prestes's office, Washburne turned to a companion and asked how much that was in dollars. Told that it was approximately half a million, Washburne sadly shook his head and said, "Poor Mr. President: he won't find oil! That won't produce anything."[21]

As Washburne was concluding his survey, Euzébio de Oliveira, still director of the SGM, commented on the oil potential of the south in his 1928 report. Where for some years he had challenged Israel C. White's pessimistic assessment, he now challenged Washburne's opposite evaluation. While admitting that his own conclusions, like Washburne's, were based upon shallow drilling, he was now pessimistic about finding oil in commerical quantities in the south. But he still considered the region "of some interest" for oil. He recommended concentration of effort on the coast from Bahia to Rio Grande do Norte and along the Peruvian and Bolivian frontiers of the territory of Acre, as the areas "most favourable for petroleum". This represented some six per cent of national territory, or about 300,000 square miles, a tremendous effort of exploration in any country.[22]

Exploration in the last half of the 1920s continued fitful. Records exist of several private companies, most of which were active in the south, but little detail of their activities has survived. Those companies that managed to reach the drilling stage struck little more than "shows" of oil, and most abandoned the effort for lack of capital. Indeed, two stopped drilling because their machinery was not powerful enough to recover the pipe and drill bit from the wells.[23]

The SGM presented no better picture, and its minimal activity was hampered by budgetary uncertainty. For example, in 1927 a shipment of drills from the United States never left port because the shippers insisted on prepayment and the Treasury Office (Tribunal de Contas) would pay only C.O.D. The material naturally did not arrive and the allotted funds were removed from the budget.[24]

As a result of Euzébio de Oliveira's influence, some Brazilians in the late 1920s urged the formulation of a national oil policy. Journalist Solidónio Leite, for example, wrote a series of articles early in 1927 in the prestigious Rio daily, *Jornal do Brasil*, demanding such a move. For Leite, petroleum was vital both for national wealth and for "the maintenance of independence" and sovereignty. He contended that in oil legislation Brazil was behind other countries. "Here everything remains to be done," he wrote. "Worse still, what we have done is precisely the opposite to what we should have done." State governors, he charged, had been permitted to make long-term contracts which gave foreigners exploration monopolies. He mentioned two such contracts, one for thirty-five years and the other for fifty, giving full and exclusive rights for entire states.* He urged that Brazil follow the "world trend" on petroleum legislation, and enact both a petroleum law and a constitutional amendment enabling the federal government to control mining concessions. Leite commended Oliveira's 1923 legislative proposal, which would have imposed some nationalistic restrictions upon oil exploration. He concluded with a plea for conservation:

> Let us follow the example of the most wary countries. Let us pass from words to deeds; from programs to their realization. Let us make our petroleum policy. Let us discuss the defence of the oil fields in the Brazilian subsoil; our preparation to slow down as much as possible their exhaustion, and their replacement when it becomes necessary.[25]

---

* While Leite did not elaborate, it was later revealed that Amazônas had awarded fifty-year exploration concessions to three foreign companies; the territory involved was all the Amazon Basin within the state. São Paulo also awarded a similar concession in the Paraná Valley to Standard Oil of New Jersey; the company did not explore but held the land in reserve.

Partly as a result of such agitation, in June 1927 the Ministry of Agriculture called upon its former Minister, Ildefonso Simões Lopes, to study the oil question and make policy recommendations. He recommended increased federal-government activity in petroleum and a special law for oil. He urged university-level training for petroleum geologists, and foreign training for all oil technicians. The federal government, he felt, should supervise all exploration contracts and investigate those contrary to the national interest. All military installations on the national frontiers should include qualified technicians for mineral and biological surveys. Finally, he urged a substantial budget increase for the Geological and Mineralogical Service.[26] Two days after his recommendations, Simões Lopes, then a Federal Deputy, introduced in the Chamber of Deputies a bill embracing his proposals. Its main provision put subsoil wealth and development both under the jurisdiction of the central government, and Article 2 read in part: "Oil fields may not belong to foreigners, nor be exploited by them."[27]

On June 30, Simões Lopes defended his bill before the Agriculture Committee of the Chamber of Deputies. The needs of national defence, he asserted, and the necessity of cheap transport forced Brazil to develop petroleum. This would ensure the growth and future prosperity of the nation as well as provide for its security. "There could not be, then, a problem which more fundamentally affects the great interests of Brazilian life," he claimed. Much subsoil wealth and hydraulic potential had been alienated to foreigners; "if we do not look clearly toward tomorrow," he continued, "dominion over petroleum similarly will pass to them." Private enterprise, he maintained, was wasteful; to increase the meagre drilling effort required government control. Since the United States was then talking of hemispheric control of petroleum, Simões Lopes called for passage of his bill to defend Brazilian sovereignty.

Article 2 sparked a great controversy because it prohibited foreign exploitation of oilfields. Simões Lopes acknowledged the need for foreign capital in national development but emphasized that "the case of petroleum has exceptional military and economic relevance"; foreign capital could not be allowed to gain control of such a vital commodity.[28] Oliveira did not, however, agree with Simões Lopes, and on July 22, 1927, he presented his own less restrictive bill. He merely wanted to bar foreigners from strategic areas. Specifically, Oliveira proposed that foreigners be allowed to operate anywhere except within sixty kilometres of the coast and the national frontiers. This bill, substantially the same as his 1923 bill, permitted exploration companies to be seventy-five per cent foreign-owned.

Oliveira had obviously become conscious of the enormous capital outlay required for oil development, and of the difficulty in raising such money in Brazil. The constitution, Oliveira noted, prohibited foreigners from "possessing minerals necessary to national security and defence". It did not prohibit them from exploring for such minerals. Without foreign assistance he believed that the projected petroleum industry would never become a reality, a view rejected by the Agriculture and Committees of the Chamber of Deputies. They believed foreign oilmen, once they had entered the country, would "easily perceive a legal stratagem" to usurp control. They therefore backed Simões Lopes's bill.[29]

The oil controversy continued in the Chamber of Deputies into 1930, while various amendments and subsitute bills were presented. The last act in the drama occurred on August 8, 1930, when Deputy Graccho Cardoso of Sergipe introduced a bill which was substantially Oliveira's 1927 proposal. Within a few months, however, as the bill was about to be passed, the 1930 revolution occurred, and Cardoso's measure died with the First Republic.[30]

By 1930, several characteristics of Brazilian oil exploration had become apparent. First, private initiatives had been without exception mounted in ignorance of both petroleum geology and capital requirements. Second, the federal and some state governments were cautiously increasing their interest, although their efforts were still minimal. Third, and most significant, there was a persistent xenophobia — distrust of foreigners — characterizing the attitude of governments toward oil exploration. Such an attitude provided fertile ground for nationalism, as would soon be seen. This nationalistic bias, finally, was based upon the assumptions that the Brazilian subsoil held substantial quantities of oil, that foreigners wanted or would want it, and that their getting it would have adverse strategic and economic effects on the country.

2
GOVERNMENT BECOMES CONCERNED, 1930–1943

In November 1930, Getúlio Vargas, defeated presidential candidate and leader of a subsequent armed insurrection, marched into Catete Palace to become Provisional President of Brazil. Vargas came from Rio Grande do Sul, a state that had emerged only recently as an economic power with prosperity based upon a mixed economy. Its politicians had decided to challenge the political dominance of São Paulo and Minas Gerais states, which had monopolized the presidency throughout the twentieth century. The higher echelons of the military had apparently concluded that the First or Old Republic had run its course, because they deposed the incumbent government before Vargas's forces could triumph in battle. Clearly, change was in store for Brazil.

The ease with which the old political system was swept aside showed its bankruptcy. It had relied for power on a rural electorate, controlled by landholders; new forces, particularly from the growing city of São Paulo, now demanded representation. The old system had relied for economic power on the export of coffee, whose price had declined sharply after the mid-1920s, until the 1929 crash had driven it into collapse. Finally, the Old Republic fell victim to the world-wide demand for political and economic change in the face of the Depression.

The Provisional Government responded to Brazil's economic and political malaise in a fashion common to other new governments of the time: centralization of political power and growing state control over the economy. Until a new constitution could be drafted, for example, all legislative bodies (from the Congress to municipal assemblies) were abolished, and the Provisional Government, through its state and local appointees, called *in-*

terventores, assumed full legislative as well as executive authority.

The Provisional Government drew on traditional Iberian paternalism and contemporary corporatism in its attitude toward the Brazilian economy, particularly toward the exploitation of natural resources. Vargas was also able to capitalize on the divisions among supporters of his *coup*—united only in their dissatisfaction with the recent governments of the Old Republic—to realize his personal inclinations toward political centralization.

The political and economic illness of the state required a drastic cure. Thus, shortly after the triumph of the revolution, the Provisional Government established a Grand Legislative Commission, which, among other things, struck three subcommittees to formulate codes for mines, water, and forests (João Pandiá Calógeras headed the first). Before the end of 1931 the federal government had overridden the 1891 Constitution and decreed its right to authorize the exploration and exploitation of mineral resources within the entire territory of the republic. Thus, both landowners and states had lost control of mines and mining.[1]

In the same year the government claimed from the foreign petroleum-product distributors $6 million for alleged back taxes.* The Council of Taxpayers, a judicial body constituted under the Provisional Government, heard the first case (against Royal Dutch–Shell) in 1932; it found in favour of the company and rejected the government's claim. Had the decision gone the other way, the Provisional Government might well have confiscated the holdings of the five distributing companies.[2]

Vargas's centralist-minded followers clearly intended that only the federal government should control mining, including petroleum. All state concessions to foreign oil explorers were cancelled, and the new government began to consider promot-

---

* The five distributors were: Standard Oil of New Jersey, since 1912, the Texas Company, since 1915, and Anglo-Mexican (later Royal Dutch–Shell), Atlantic Refining, and the Caloric Company (for fuel oil only), all since 1921. All imported finished products for resale; none operated refineries in Brazil at that time.

ing oil exploration by the SGM as well as assisting private prospectors. Early in 1932, for example, General Assis Brasil, Vargas's first Minister of Agriculture, hired American geophysicist Mark Malamphy "to blaze new trails in the search for oil".[3]

Government and private industry, however, were soon at odds over petroleum exploration. Late in 1932 a young engineer, Manoel Ignácio Bastos, reported an oil seepage at Lobato, near Salvador, Bahia, to the Ministry of Agriculture. Bastos, aware of long-standing reports of surface petroleum around Lobato, had dug a well and found oil seeping into it at shallow depth. Ministry technicians refused to believe him; indeed, some officials charged he had bought oil and poured it into his well. Frustrated in his attempts to solicit government support, Bastos turned to Oscar Cordeiro, president of the Bolsa de Mercadorias (Merchandise Mart) of Bahia.

Cordeiro also failed to secure government support. The Ministry of Agriculture stood by a 1932 survey which showed the geology of Lobato to be unsuitable for petroleum. Cordeiro actually sent a sample of Lobato petroleum to the Central Laboratory of Mineral Production, which concluded that only a study of the local geology could determine whether the sample was in a natural state. The Ministry of Agriculture, however, was satisfied that it had already made such a study, and refused to consider Cordeiro's request for assistance.

While the Ministry's treatment of Bastos and Cordeiro was regrettable — they were later vindicated when oil was found beside their crude well—it can be explained. On the one hand, the SGM and the bureaucrats of the Ministry responsible for oil exploration were unusually insular and intolerant to outside pressures or ideas, the result of Euzébio de Oliveira's influence. On the other hand, Cordeiro had a reputation as a "booster" for Bahia, and was not a geologist. Although Bastos had better credentials than Cordeiro—he could hardly have had worse—he apparently lost interest in the project almost immediately after contacting Cordeiro and turned his hand to other concerns.

Cordeiro, however, was incensed at the rebuff the Ministry had handed him and at the insinuations some SGM geologists had made about his methods. He launched a campaign in the press, reiterating his claims for Lobato and damning the SGM. He also badgered individual geologists for years, in the hope that someone would properly investigate his claim. As we shall see, his efforts finally bore fruit.[4]

In 1933 Juarez Távora, a leading figure in the 1930 uprising, became Minister of Agriculture. He was a strong centralist who believed in measures such as the proposed Mining Code as instruments to effect a more just, nationwide distribution of the fruits of natural-resources exploitation. He began by reorganizing the bureaucratic apparatus concerned with oil exploration. In July 1933 he formed the General Directory of Mineral Production (DGPM), which in March of the following year became the National Directory of Mineral Production (DNPM). To it Távora transferred all mineral properties, prospecting permits, and mining concessions which had been under the jurisdiction of the states. Under the DNPM he placed the SGM and another new agency, the Service for the Stimulation of Mineral Production (SFPM). Unfortunately, the reorganization was not accompanied by a budget increase; the government continued to appropriate meagre sums for oil exploration.

To the SFPM flocked Brazil's brightest young petroleum geologists: men such as Domingos Fleury da Rocha, Glycon de Paiva, and Irnack Carvalho do Amaral. They would subsequently become the most famous oilmen in Brazil, and have been intimately involved with the industry up to the present day. Joining them were the American geophysicist Mark Malamphy, and a young technician of Lithuanian origin, Victor Oppenheim, who had come from the Argentine state-oil entity. Oppenheim had had no training in petroleum geology and little geological experience upon his arrival in Brazil. Unfortunately he was too vain to acknowledge this deficiency and he learned much of his chosen calling on the job. His inexperience eventually had disturbing consequences.[5]

The private sector, which had lagged somewhat at the outset of the new political regime, signalled its revival in April 1932, when an electrical engineer, Edson de Carvalho, was authorized to explore for oil. Carvalho had earlier been given concessions in the Riacho Doce region of Alagoas, an area with good surface indications for oil. Soon he announced the organization of the Companhia Petróleo Nacional, with an initial capital of 20,000 contos (*c*. $2 million), wholly financed within Brazil; half was in rights and concessions in Alagoas and the remainder was in shares sold to the public.

One of the directors of the company was José Bento Monteiro Lobato, well known in Brazil for his children's stories; he was also a publicist and an ardent nationalist. He appears to have been relatively ineffective in eliciting support for his new company, however; shares did not sell and the company had lost half its capital by the following year. To direct operations the company contracted with Señor F. B. Romero, a Mexican, who had a mysterious apparatus that located oil without drilling. Apparently it was a galvanometer hooked up to two bars of metal of different kinds which were stuck into the ground. A difference in potential between the two bars — and thus an electric current — resulted. By harnessing the delicate current generated and using it to move a needle across a graduated scale, the operator could demonstrate to his fascinated observers just how many thousands or millions of barrels of oil were just beneath the surface.*

To carry out its operation—since in fact the company had no cash—it borrowed, from the Alagoas government, a drilling rig equipped for geophysical exploration which the SGM had originally loaned the state. Henrique Lage, a wealthy Brazilian investor, purchased a new rig for the company, accepting payment in shares. The engineer directing drilling resigned almost immediately, so Carvalho appealed to the SGM. Mark Malamphy recommended Victor Oppenheim, who had just arrived in Brazil and was looking for work. In January 1933 Oppenheim began drilling in Riacho Doce; within a month he had reported to Carvalho that sediments in the region were only 500 metres deep, and beneath them was crystalline rock. There were no prospects of oil there. Oppenheim thereupon left the company, went to Rio, and joined the SFPM. Romero's apparatus, however, had already discovered "great sheets of oil" in the same area, on the basis of which the company had presented a stridently optimistic report to President Vargas, and had re-offered its shares to the public; so, after the company's exciting announcement, a Rio newspaper interviewed Euzébio de Oliveira, still

---

* Mr. Barnum, I know what you're going to say!

director of the SGM. Oliveira said not only that he doubted the existence of significant quantities of oil in Riacho Doce, but that he questioned both the reliability of Romero's apparatus and the integrity of the Companhia Petróleo Nacional directors.

Such a negative statement from an influential figure had a devastating effect on the company.[6] Quite simply, it could not sell a share, but Carvalho carried on in Alagoas with his borrowed equipment and his own and borrowed funds. In September he reported "contact" with oil, at "minimum depth" and "only 600 metres from the seaport". Nothing further was heard for two years, and Carvalho's claim seems to have made no impression on the rest of the country.

When Carvalho reported his strike, Monteiro Lobato decided to make an issue of what he regarded as sabotage on the part of the SGM. He fired off a heated letter to Vargas, denouncing the SGM and Oliveira as saboteurs, repeated the charge in the introduction of a book on oil he translated, and began to campaign in the press to discredit the DNPM and its agencies.[7] This precipitated a bitter and hysterical feud between government technicians and private oil speculators, a feud carried on in the public press and with frequent disregard for truth.

Monteiro Lobato, despite his championing of the Companhia Petróleo Nacional in the press, left the company very soon after Oliveira's denunciation. He went south to São Paulo and formed the Companhia Petróleos do Brasil, with an initial capital of 3,000 contos (c. $300,000). It was to drill at Araquá in the *município* of São Pedro, an area which, among others, Euzébio de Oliveira and Chester Washburne had considered favourable.

By a bizarre coincidence, the SFPM sent Victor Oppenheim to the same area at this time. He again disputed majority opinion among Brazil's geologists and pronounced the region "frankly negative" as an oil prospect. He further recommended that the SFPM direct its attention away from the coast to the Andean periphery, particularly the territory of Acre. The agency did in fact divert some of its meagre resources to Acre and in 1934 sent a party under geologist Pedro de Moura which remained there until 1940.

Monteiro Lobato was unaware of Oppenheim's assessment, however, and based his optimism on Washburne's earlier research. With the São Paulo government's collaboration, he also acquired a large drilling rig and announced that the company would drill a hole twenty-four inches in diameter and 2,000 metres in depth (this was over 1,500 metres deeper than the SGM had reached in any of its twenty-two wells). None other than Señor Romero was directing operations, however, and he had chosen the site with the aid of his remarkable appartus. Monteiro Lobato was more successful in publicizing this company than his last; the initial capital was quickly raised, and most of it appears to have come from small investors.[8]

The emergence of Monteiro Lobato's new company coincided with a rash of rumours about the secret activities of the international petroleum companies, based on the allegation that representatives of "the trusts" had secretly surveyed the sedimentary basins of Brazil, drilled wells, and capped them for future use. No one had found the wells, which made the rumours appear all the more sinister. "The trusts" were also said to be buying up likely areas of the country through Brazilians who were figureheads (*testas de ferro*). One rumour, which Monteiro Lobato circulated, maintained that foreign oil companies were trying to convince Brazilians that their country had no oil so that the companies could buy up rich areas without suspicion. They were also said to be furthering these rumours by suggesting that if Brazil had oil, Americans would already have taken it out.[9]

Drilling began at Araquá with great publicity. When the well had reached 400 metres, Señor Romero announced that the discovery of oil was imminent. Several newspapers used the occasion to comment on the lack of effort from the government, whose only function seemed to be to disparage genuine attempts to bring benefits to the country.[10] Monteiro Lobato halted drilling to solicit more capital and said in an interview, "The problem of petroleum is practically solved." He belittled the DNPM effort and accused the government agency of believing in "the existence of petroleum in all the countries of the Americas, *except* the largest of them".[11]

As the months went by and the Companhia Petróleos do Brasil did not find oil, however, Monteiro Lobato's optimism gradually faded and public recrimination against him began to appear. By January 1934 Rio's *Jornal do Brasil* was recommending "a little care" on the part of the government about petroleum, bearing in mind Monteiro Lobato's apparent failure. The newspaper charged that it was the responsibility of the government to prevent such "constant organizers of 'petroleum companies' " from preying upon the citizenry.[12]

As if in response to the charge made in the *Jornal do Brasil*, the Vargas government promulgated a constitution and a mining code within a week of each other in July 1934. Each increased the control the central government exercised over oil exploration and exploitation. The new constitution included, as Articles 118 and 119, the 1931 decree which had separated ownership of the soil and the subsoil and had given the central government the exclusive right to grant concessions for subsoil exploration and exploitation. The latter article also warned that mines, mineral deposits, and waterfalls would be "gradually nationalized". The new Mining Code reiterated these provisions and set out requirements for the granting of concessions, which included a plan of attack submitted with the application to the Ministry of Agriculture, which would consult the DNPM.[13]

There were clearly several impulses which came to bear on the two documents as they related to mining. The *tenentes* ("lieutenants") — the reform-minded, corporatist young wing of the army — strove for development within strict limits, for tying productive sectors and energy sources to the central government to bring about a more equitable distribution of wealth (which, as Juarez Távora claimed, would come about only under centralized controls). The army as a whole was beginning to worry about oil for security reasons: it had become concerned about a possible foreign takeover of a future oil industry, and it was not satisifed that national private capital was sufficient to establish such an industry without borrowing from outside the country (with a consequent loss of control over the resource). Prominent civilians, such as João Pandiá Calógeras, who had helped draft the code, clearly agreed with the military on this problem.

The government reorganized and revitalized the DNPM, which began a systematic study of the petroleum possiblities of the country. One of the first areas it surveyed was Araquá, where the Companhia Petróleos do Brasil was still drilling; the government crew soon abandoned the region, reporting it "frankly negative" for petroleum since the basin was structurally a graben.*

The conflict between two different types of nationalism was becoming increasingly bitter. Each side — government and private industry — wanted to exclude foreigners from petroleum exploration, but industry was willing to gamble its money and asked others to share the risk, while government tried to protect citizens from being victimized by overenthusiastic private companies. The government's role as protector was ineffective, however, in the absence of greater effort to supplement private initiative in oil exploration. Indeed, the hostile attitude of government geologists toward private-company exploration probably retarded exploration.

As the controversy grew more heated, the Companhia Petróleos do Brasil found it needed operating capital. In October 1934 it published, in the newspaper *O Estado de São Paulo*, a "Manifesto for an Increase in Capital". The manifesto claimed that the company's well at Araquá was outside the graben that the DNPM had identified, since drilling had not encountered diabase to 1,044 metres. It further claimed that Fleury da Rocha, Director of the SFPM, had at one time criticized the work of government exploration teams as "totally inadequate", since none of their wells had reached 800 metres.

The DNPM replied immediately in "all newspapers and telegraph agencies", asserting that its geologists had seen no oil in the well of the Companhia Petróleos do Brasil at the level at which the company claimed it had found traces of petroleum. The DNPM reiterated that it still saw no reason for optimism in the Araquá area, but condescendingly stated that it did not deny the stratigraphic and geologic value of continuing work in the area.

The Companhia Petróleos do Brasil reacted with a savage attack on the intentions of government geologists in petroleum exploration. It charged that "a group of creatures" existed in the DNPM whose motto was "NOT TO EXTRACT PETROLEUM AND NOT TO LET IT BE EXTRACTED". The company added that it was the victim of

---

* A depression in the earth formed by sinking, not by faults in the rock, and thus unlikely to trap oil.

"systematic opposition and moral sabotage, by the press [and] by the telegraph agencies". The intent of the DNPM, the company declared, was to keep Brazil "THE SOLE COUNTRY OF AMERICA WITHOUT PETROLEUM". The article concluded:

> If there is no petroleum, why stop us convincing ourselves . . . at our own cost? Why try to hold us back, if the result will necessarily be negative? What fear, what terror, what worry is this that the Araquá well inspires? Can it be that the Geological Service is not absolutely convinced of its affirmation and worries that our well will at any moment put the lie to their dogma that there is no petroleum in São Pedro—and are not many people examining this "graben"? It is useless for them to proceed with their campaign. The Cia. Petróleos is sure in its course and nothing will dissuade it from proceeding to the finish and from giving petroleum to Brazil.

The company succeeded in raising new capital (the controversy certainly had done it no harm), and work proceeded on its well at Araquá. Fleury da Rocha, Director of the SFPM, threatened Monteiro Lobato with "Justice", and called for an inquiry commission. According to Monteiro Lobato's biographer, that was exactly what he wanted. Indeed, when a commission was not immediately convened, Monteiro Lobato repeated his own charges in letters to the DNPM, adding that the government agency was falsifying reports to dissuade national oil companies, hoping to keep Brazil an importer of petroleum. Monteiro Lobato asserted that the DNPM was an agent of the conspiracy organized by the foreign petroleum "trusts".[14]

Meanwhile, early in 1934 the DNPM had sent geologist Victor Oppenheim to Lobato in response to the "continual and never-dampened insistence" of Oscar Cordeiro. Oppenheim visited the site of Bastos's well and concluded that the petroleum (which Bastos claimed to have discovered in 1931) was not natural to the locality, but had been put there. It was "ridiculous to speak of oil in Lobato", said Oppenheim, since the sediments were shallow there. (He did not order drilling, but reached his conclusion because the oil was found only a few metres from crystalline structures.)[15]

Cordeiro kept trying to enlist the support of the DNPM to drill a proper oil well near Bastos's cistern. Finally Juarez Távora, Minister of Agriculture, wrote Cordeiro what was intended as a final answer on May 14, 1934:

> In response to your letter of the 8th instant, I inform you that the opinion of the geologist Victor Copenheim [sic] is the opinion of the technicians of the National Department of Mineral Production and may be summarized as it is in the official letter sent to you by the Director of this Department. This ministry cannot have an opinion which is not that of its technicians, who have studied sufficiently the matter which interests you.[16]

Monteiro Lobato then entered the dispute, charging in an open letter to Távora that Oppenheim had come "directly from the trust which has a program to conserve Brazil in a state of petroleum slavery."[17]

Cordeiro's campaign did win one convert with credentials as a geologist. Sílvio Fróes Abreu, teaching at the National Institute of Technology, received a sample of Lobato oil, analysed it, and was sufficiently intrigued to go to Lobato. He concluded that the area was worth investigating, but was powerless to do anything on his own; he was also without influence with the DNPM, which had obviously made up its mind about Lobato.[18] Within a few years, however, Fróes Abreu would be instrumental in having the SFPM drill at Lobato, and with dramatic results.

In mid-1935, as the dispute in the press between the DNPM and Monteiro Lobato got more and more heated, Edson de Carvalho had the Governor of Alagoas wire the new Minister of Agriculture, Odilon Braga, that he had struck gas in Riacho Doce at a depth of 262 metres. The Governor requested that the SFPM verify the finding, which it did and recommended proceeding with exploration. Alagoas then hired a German geophysical firm, but the SFPM also began geophysical exploration in the same area. The result was a tremendous outcry from Carvalho and Alagoans, charging that the SFPM was trying to undercut the German firm and distract attention from its work. To make matters worse, the German firm pronounced Riacho Doce "absolutely oil-bearing", while both Malamphy and Oppenheim wrote a joint refutation of its report. In the end the Companhia Petróleo Nacional collapsed, but the issue remained bitterly deadlocked.[19]

The controversy in the press over the merits of government and private industry intensified throughout 1934 and 1935. It became so bitter that in March 1936 the Minister of Agriculture, Odilon Braga, persuaded President Vargas to constitute a full official inquiry into the issue. Braga felt the controversy had stirred such emotion because Brazilians could not accept that so vast and rich a country did not hold potent oil resources, since petroleum was generously distributed throughout most of the bordering republics. Local financiers were not so confident, and yet small companies began with a minimum of capital because their organizers believed little effort would be needed to find oil.[20]

In the meantime, not all government geologists were as sceptical of Oscar Cordeiro's claims about Lobato as Oppenheim had been. Three young members of the DNPM, Glycon de Paiva, Irnack Carvalho do Amaral, and Sílvio Fróes Abreu (who had joined the DNPM since visiting Lobato) decided in 1936 to make a thorough survey of the Bahia Recôncavo, inspired, as Fróes Abreu later wrote, "by the oil seepage of Lobato and by the history of rather suggestive surface indications, confirmed by persons of complete integrity". The three were able to secure financing for the venture from Guilherme Guinle, a prominent Rio banker.[21]

Despite "a concept almost fixed [by] the national body of technicians, . . . that the petroleum of Lobato 'was Cordeiro's obsession' ", they concluded that oil was indeed seeping into the well dug by Manoel Ignácio Bastos, and Fróes Abreu so informed President Vargas, the directors of the DNPM and the SFPM, and the Petroleum Inquiry Commission. The geophysical studies the young geologists had carried out had indicated the true depth of the basin, had proven "the generalized occurrence of schists, with organic material capable of becoming oils or gases, and proved moreover the existence of tectonics favourable to the accumulation of oil."

They then published a book in 1936, *Contribuicões para a Geologia do Petróleo no Recôncavo, Baía*, predicting that drilling would encounter commercial deposits of petroleum in the Recôncavo. Their purpose, Fróes Abreu later wrote, was to force the area to official attention. Only one member of the Academia Brasileira de Ciências, Mathias G. Roxo, commented on the book and he expressed doubt that the sediments were of significant depth.[22]

Fróes Abreu also testified before the Petroleum Inquiry Commission in 1936 concerning the perennial charge that "foreign interests" had impeded the discovery of oil. He stated that after careful and patient study "we have come to the conclusion that there has never been any evidence of foreign sabotage." Brazilian working conditions exhibited so many shortcomings that foreign groups, "well acquainted with the progress of our exploration, did not worry about Brazilian competition in their oil market, cordially shared by the companies here."[23]

This charge originated with Monteiro Lobato, who, in his 1936 book *O Escândalo do Petróleo* (*The Petroleum Scandal*), charged that "the trusts" had set up figurehead companies to hold back the national exploration effort. Fróes Abreu declared he knew of two short-lived Brazilian companies linked to international consortia: Companhia Geral de Petróleo Pan-Brasileira (Standard Oil of New Jersey) and Companhia Brasileira de Petróleo (Royal Dutch–Shell). The companies had been set up for exploration purposes, and had existed before 1934 when there was no formal legal impediment to such activity by foreigners. [24]

The Petroleum Inquiry Commission fizzled away in 1937 after issuing a blanket exoneration of the DNPM and its methods. While the Commission did not condemn Monteiro Lobato for his activities, it seems quite clear that Monteiro Lobato was trying to inflate the value of his stock through the publicity the inquiry would give him. It is equally clear that he misled the Brazilian public about his investments — indeed, the beloved author was nothing more than a confidence man—and used the role of an injured patriot to good effect. Yet the DNPM technicians, because of their arrogance and intransigence, must share the blame for the episode.

Despite the fact that Monteiro Lobato never found any oil, he left a legacy to Brazilian petroleum which has not been forgotten. The first part of the legacy was epitomized in his book *The Petroleum Scandal*, which, because of his fame, was widely read and quickened the Brazilian public's interest in oil as nothing else did. The remainder of the legacy involves style: principally due to Monteiro Lobato, rhetoric about petroleum became more combative and less concerned with a strict adherence to facts. Monteiro Lobato, and those who followed in his footsteps, tended to make light of the complicated technical and financial aspects of the oil industry and focus on conspiratorial politics. By the mid-1930s, it is important to note, the little technical information about Brazil's petroleum geology that had been published was very tentative and found in official reports or learned journals; it was thus virtually unavailable to the overwhelming majority of those who read Monteiro Lobato's books, and they were unequipped to criticize his rhetoric.

The most significant development in Brazil during 1937 was a *coup* from within the government, by which Vargas was able to establish a corporatist dictatorship, the Estado Nôvo, a not uncommon sort of regime in those troubled times. In this move, the President relied on the support of a like-minded wing of the military which had, as one of its most experienced and respected members, General Júlio Caetano Horta Barbosa. An engineer who had served his country well on difficult projects in its interior, Horta Barbosa was deeply patriotic and filled with a sense of the army's role of leading the nation toward integration and development. He was to play a major role in future oil policy.[25]

The constitution of the Estado Nôvo, enacted in the same year as the *coup*, was even more nationalistic and centralistic than its 1934 predecessor. It stipulated, for example, that only Brazilians could own shares in national mining or petroleum companies; no foreign capital could participate, nor could national capital belonging to foreigners residing and "enriching themselves" in Brazil.[26]

Before the new constitution came down, the Vargas government had taken the first steps toward creating a national oil industry: it altered the tariff structure to attract crude and fuel oil, rather than such more refined products as gasoline and lubricating oils, and to encourage their processing within Brazil. As a result, between 1935 and 1937 twenty-five diesel-distillation plants appeared, at least four of which also handled crude oil and thus could be called refineries.[27]

Horta Barbosa was the chief proponent of such a move, but it was to be not an end but a beginning in his mind. As a good soldier, he was concerned for Brazil's security without crude-oil reserves of its own; he hoped that a refining industry would finance exploration for his nation's undoubted oil. He was concerned, however, lest such an effort fall into the hands of "the trusts", and he doubted both that Brazilian private companies could ever command the necessary financial and technical resources to find oil, and that they would necessarily operate in the best interest of Brazil; he therefore urged that the army be given the task.[28]

At last, in 1938, the Vargas government made a number of moves which appeared to indicate a genuine interest in exploring for oil, moves which indeed would shortly bear fruit. First, Brazil negotiated an agreement with Bolivia by which it would complete a long-promised link to the sea for Bolivia (at this time by rail to Santos), in return for which Bolivia would set aside part of its apparently oil-rich "subandean" region for joint exploration and exploitation with Brazil. Further, the agreement specified that after Bolivia's tiny consumption requirements had been satisfied, the surplus oil would belong to Brazil without qualification.[29] Then, two of the DNPM's brightest young technicians, geologist Glycon de Paiva and geophysicist Irnack Carvalho do Amaral, began actively to promote a radical change in exploration policy: the hiring of foreign drilling specialists. Agriculture Minister Fernando Costa overrode the objections of DNPM administrators and called upon the two geologists to investigate such possibilities and draw up a three-year drilling program under their new guidelines.

In publishing their ensuing recommendations, Paiva and Amaral offered some restrained criticism of government efforts in exploration to that time. The SGM had begun its oil exploration in 1918 with a much too scientific and theoretical attitude, they said, because it employed no engineers, only geologists. No real exploration could have been made in such circumstances. They regretted that the Ministry of Agriculture, under whose jurisdiction oil and minerals exploration lay, was the poorest of all the ministries; this was hardly conducive to the growth of an industry requiring, by its very nature, vast funds and contact with the outside world.[30]

The Estado Nôvo was to be much more dynamic, as well as much more authoritarian, than the regime from which it sprang. In 1938, shortly after the November *coup* which brought it to power, the new Vargas government enacted several decrees which underscored the nationalism of he 1937 Constitution. Three of the decrees pertained directly to petroleum, and reflected both Horta Barbosa's wish that the government should control the oil industry and that of Paiva and Amaral that the government oil effort should be revived and directed. Decree-Law 366 declared all oilfields yet to be discovered within the national territory to be the property of the federal government. Decree-Law 395 declared the national supply of petroleum a public utility, nationalized the refining industry, and created the National Petroleum Council (CNP) to control the industry. Decree-Law 538 gave structure to the CNP, fixed its prerogatives, and laid down the policy it was to pursue (see Appendix I). Its purposes were to stimulate a national refining industry capable of supplying the country's needs and to carry on a systematic search for oil which, if found, likewise would be exploited under national control. The CNP would control exploration, but was not barred from letting out contracts for such activity.[31]

Events outside Brazil also influenced the government in creating the CNP. Possibly the Mexican move of March 18, 1938, expropriating the entire petroleum industry, influenced Vargas and his advisers; the CNP came into existence very soon thereafter (on April 29).[32] Moreover, the Mexican experience *before* nationalization may have influenced Brazilian thinking, prompting restrictive legislation before the industry had come into existence. Brazil obviously hoped to avoid Mexico's unhappy experience in trying to regulate an already prosperous business.

It was the Argentine government oil agency, Yacimientos Petrolíferos Fiscales (YPF), however, that served as a model for the "structure and . . . spirit" of the CNP. Formed in 1922 just *after* the initial discovery of oil in Argentina, YPF was created to regulate the national petroleum industry as well as to engage directly in it. The CNP came into being *before* the discovery of oil in Brazil, but with the same role.[33] Yet for Monteiro Lobato, the creation of the CNP was ridiculous. "Since there is no petroleum in Brazil," he said, "as the government affirms, for what purpose is a council for that which does not exist?" He charged that his Companhia Petróleos do Brasil had had to abandon its Araquá well "because of legislation", and that the CNP had forced the Companhia Petróleo Nacional to give up its work in Alagoas.[34]

Another event that took place in 1938 was the formation, prior to the promulgation of Decree-Law 366, of a company to explore the Lobato region of the Bahia Recôncavo. The organizers were the Rio banker Guilherme Guinle, who had sponsored the 1936 search there by Sílvio Fróes Abreu, Irnack Carvalho do Amaral, and Glycon de Paiva, and the Paulista bankers Murray–Simonsen. To carry out the exploration, the new company contracted with J. E. Brantly of the American firm, the Drilling and Exploration Company. The subsequent decrees nationalizing exploration, however, forced the company to dissolve.[35]

Fróes Abreu and Paiva, perhaps upset at the dissolution of this company and exasperated by the DNPM's refusal to reconsider its position on Lobato, then published outspoken criticism of Brazil's oil exploration up to 1938. In the *Revista Brasileira de Geografia* late in 1938, Sílvio Fróes Abreu painted a dismal picture of exploration. He wrote that the previous twenty years had demonstrated three theses: that public power was not adapted to petroleum exploration; that Brazilian capital was not interested in the activity; and that foreign groups were not anxious to find petroleum in Brazil. Lacking funds, the government could not pay enough to attract the best men, and used those it had infrequently because it could not afford to keep its crews in constant operation. Nationalistic legislation, moreover, offered no real incentive to explore for oil. Oil had not been discovered in Brazil, concluded Fróes Abreu, because there was no "stimulating legislation".[36] Glycon de Paiva summed up the exploration effort in April 1938: "Up to now we have pretended to look for petroleum in the national subsoil."[37]

The latter assertion was not far from the truth. The geological department of the federal government had drilled only sixty-seven wells from 1919 to 1938. Half had been sunk to depths of less than 300 metres, and the deepest well, 768 metres, had taken two years and four months to drill.[38] In 1938, however, the SFPM, bowing to pressure from its younger members and from (still) Oscar Cordeiro, did at last send a drilling crew to Lobato. On the advice of one of his geologists, SFPM Director Avelino Ignácio de Oliveira ordered a well drilled near the site of Bastos's cistern "to sufficient depth to settle the question once and for all". The rig first struck crystalline rock at 22 metres, and then encountered oil-impregnated arenite at 71 metres in a second hole. Oliveira ordered a bigger rig transferred from Paraná to drill in a different location.[39]

While this was taking place, Sílvio Fróes Abreu, in December 1938, was received into the Brazilian Academy of Sciences. The work he presented upon induction was *Contribuições para a Geologia do Petróleo no Recôncavo, Baía*. The members of the Academy, Brazil's elder geologists, hotly disputed the conclusions of their young colleague; indeed, one geologist declared it was a "joke" to talk of petroleum in the Recôncavo. Fróes Abreu, humiliated, put the book under his arm and left the room.[40] But just one month later the rig brought from Paraná struck oil-impregnated arenite at 214 metres. Drilling was stopped and the oil left to accumulate over a weekend, and then on January 21, 1939, Brazilian oil was brought to the surface at Lobato.[41]

The implications of the discovery were enormous. Brazil now had oil, as every Brazilian in his heart of hearts had always believed; it was surely only a question of time before consumption requirements could be satisfied with national production. Monteiro Lobato had on the one hand been vindicated: he had told Brazilians to believe in their country's oil wealth. On the other hand, however, he had been legislated out of the continuing search, along with other private companies. The timing of the discovery could hardly have been improved upon: the well had been brought in by a government agency within a year of the creation of the CNP, the government organ that was to instill vigour and direction into the search for oil in Brazil.

If any possibility of wildcat exploration remained, the CNP quickly removed it by having Decree-Law 3701 issued on February 8, 1939: all petroleum deposits in the Recôncavo, within a radius of 60 kilometres of Well 163 (the discovery well), were made a national reserve; exploration of the entire area was reserved to the CNP.[42] Moreover, the CNP refused to renew the charters of the few private exploration companies left in the country; in almost all cases, the reason given was that the company had foreign shareholders, a situation which was now illegal.

The driving force behind the zeal of the CNP to crush its competition was clearly its first president, Horta Barbosa. The

Brazilian oil industry was going to be safe from foreign encroachment, a source of security in peace and war. Apart from the obvious logic of putting Horta Barbosa into the position — given his corporatist leanings and strong advocacy of just such an approach to oil development — the choice was wise politically. Horta Barbosa had been important in the establishment of the Estado Nôvo just over a year before; Vargas gave him the position and a free hand in order to keep Horta Barbosa's military clique behind him as President of Brazil.

In April 1939, two months before the CNP began operation, Horta Barbosa visited oil industries in Uruguay and Argentina. He was impressed with ANCAP (National Administration of Fuel, Alcohol and Cement), the Uruguayan state-monopoly company for crude-oil refining, and YPF in Argentina. His tour prompted him to recommend that as many technicians as possible be sent to the La Plata countries to study the specialties of the oil industry; that the CNP control all activities relating to the industry; that advisers be solicited from Argentina; that refining be made a state monopoly; and that more funds be allotted to the CNP.[43] The rights to oil exploration and exploitation and the corresponding equipment were then transferred from the DNPM and the Ministry of Agriculture to the CNP.[44]

One of Horta Barbosa's first acts as president of the CNP was to place orders in the United States amounting to ten million milreis (less than $600,000) for "a complete machinery outfit" to explore and drill in Bahia. American oil experts were to train Brazilians to operate the machines, which were to be in use twenty-four hours a day.[45] The CNP also announced it would contract with specialized drilling firms to inject new life into the exploration effort.[46]

An act of equal significance was the submission to Vargas of two proposed decrees in July 1939, to set up a single federal tax on petroleum derivatives to replace the welter of state and local imports, and to establish a government refinery. As John Wirth, in his study *The Politics of Brazilian Nationalism*, points out, the unified federal tax was an important achievement. A uniform tax rate was set for the entire country, based on the average of all the old taxes; this enabled the CNP to set uniform prices, which lightened the bookkeeping load of the distributing companies, lowered prices in the interior, and maintained the income for the government. States and municipalities allotted 26 per cent of the fund for highway building. Two corporatist objectives were promoted: disciplining of prices and national integration.[47]

The decree establishing a national refinery had as its inspiration Horta Barbosa's recent trip to Uruguay and Argentina. Officials in both countries had stressed the importance of using the potentially high profits from refining for the good of the state in general, and the fledgling oil industry in particular: the profits were not to go to private individuals. Horta Barbosa wanted to use refining to finance exploration—the only way, he claimed, that a Brazilian oil industry would develop. The decree also sprang from Horta Barbosa's fear that to allow private developers to participate in the oil industry would be to permit the entry of foreign monopoly companies; indeed, some of the proposals for refineries which began to appear at this time used funds from just such companies.

Finally, the decree was appropriate in the light of Brazil's priorities by 1939. The government was concentrating its attention on building a steel mill, which would swallow more and more money in the immediate future. If oil exploration could be financed from within its own industry, so much the better.[48]

Accordingly, Horta Barbosa pressed the administration into several actions which would increase CNP control over oil to the detriment of private industry. First, he had the taxes on locally refined oil raised so that the profit margin from refining shrank to about 25 per cent. This change caused, for example, one consortium to abandon its plans: the governments of São Paulo, Rio de Janeiro, and Bahia states had negotiated with leading local financiers to build refineries in the three states.[49] Then, the CNP served notice on foreign oil companies that they would not be allowed to extend their activities beyond distribution. A refinery built by Standard Oil of New Jersey at Jaguaré, São Paulo, had been completed in 1938, just before the promulgation of Decree-Law 395. The CNP claimed that the refinery was "small, poor and improvised" (it had a capacity of about 2,000 b.p.d.), with no other function but "to create a precedent and compromise the effectiveness of the national petroleum policy". Whether or not such was the case, quite a different precedent resulted: the CNP prevented the refinery from going into operation, and refining became a strictly national industry.[50]

In 1940 the journal *World Petroleum* reported that the Brazilian government was "proceeding cautiously with a drilling program that appears to be yielding good results". While suggesting that development "might have proceeded more rapidly had private capital from abroad been permitted to engage in exploration", the journal acknowledged that results seemed to have justified "the fairly stable program defined by the government". Rumours persisted that "at least two wells . . . with an initial production in excess of 200 barrels per day" had been drilled.[51] This joint effort between the government and private enterprise, making as much use of foreign experience as nationalistic sentiment would allow, was a positive step toward effective oil exploration. For example, before 1939 it had taken years to drill wells of a few hundred metres; now the Drilling and Exploration Company sank a well of 1,500 metres, its second in Brazil, in three months.[52]

The CNP restricted its exploration to the accessible Atlantic Coast region in the states of Bahia, Sergipe, and Alagoas. While geologists considered the territory of Acre and the Amazon Basin promising, the dense jungle made exploration almost impossible. With the small CNP budget (equal to $250,000 in 1939, and $1,500,000 annually from 1940 to 1943, to cover *all* operations), it made economic sense to concentrate resources in the most attractive areas.[53]

In 1940 President Vargas decreed a new Code of Mines to consolidate existing laws. The code reiterated that only the federal government could authorize exploration on public or private property, and to Brazilian companies (composed of Brazilian members and stockholders) alone. The CNP was in charge of granting such authorizations.[54] Yet, perhaps prompted by the growing internal market, the discovery of oil at Lobato, or a combination of both, in 1940 "a foreign group" (apparently Standard Oil of New Jersey through its subsidiary, Standard Oil of Brazil) suggested in a confidential memorandum to the Cabinet that it might enter into exploration "if a sound and stable juridical basis could be established".

This group proposed the organization of a mixed company of foreign and national capital, the former comprising more than half. The company would explore, drill wells, and produce petroleum; it would pay all expenses and provide technicians. In return it would receive from the Brazilian government a percentage of the crude oil obtained, "with the right to dispose freely of it, as of its property". The government could either retain its share of the oil or sell it to the company at the going rate. The company would have the exclusive right to explore designated (large) areas over long periods; when agreed exploration periods expired in each region, the company would indicate which parts it wanted to be set aside for development. The company would immediately start drilling, and if oil in commercial quantities were found, it would build all facilities for proper exploitation. In addition, the company would be authorized to build refineries to supply the internal market and, once this market had been satisfied, to export petroleum derivatives. The company would pay taxes, but only to a predetermined percentage based upon the production of crude at the well mouth.

Should the contract be rescinded by the government, the company would be recompensed to its share of the calculated value of the petroleum *obtainable* from all areas developed by the company under the contract. The company also asked for freight concessions and for guarantees that transport would be maintained. It emphasized that such a plan would permit the participation of experienced oil concerns in the development of the Brazilian petroleum industry, and that "the advantages from this for Brazil are so obvious that they need not be mentioned."[55]

General Horta Barbosa, alarmed at such a prospect, sent the memorandum with his comments to General Góes Monteiro, a strong nationalist and the Army Chief of Staff. Horta Barbosa cited the Constitution, the Code of Mines, and Decree-Law 395, all of which specifically excluded foreign companies from the Brazilian oil industry, and asserted:

> The recommended solution is nothing more than that in effect in Venezuela, where the companies, due to the predominance of liquid combustibles, exercise absolute control over the complete economic life of the nation, to the extent of fixing the exchange rate of its currency.

This degree of control he termed "colonial".[56] The General Staff recommended rejection of the proposal. President Vargas's

Cabinet, however, almost unanimously approved it. Horta Barbosa accordingly offered his resignation, but Vargas, clearly alarmed by Horta Barbosa's move, assured him that the proposal would never receive executive approval. The Cabinet subsequently tabled the memorandum.[57]

In 1941 Vargas decreed a separate petroleum law, the first such legislation in Brazilian history. It provided that all oil and natural-gas deposits belonged to the federal government, with no mention of the rights of the landowner (a further increase in the power of the federal government). The CNP thereafter would grant authorizations for exploration and exploitation, and thus exercise jurisdiction over all petroleum activity in Brazil. The federal government, through the CNP, could participate in any phase of the oil industry.[58]

While the Drilling and Exploration Company, under the supervision of the CNP, drilled the majority of wells, three private Brazilian firms were also engaged in exploration under CNP authorization by late 1941. They employed Brazilian geologists and hired foreign technicians and equipment on a contract basis. Two of the companies were drilling in the state of Bahia, outside the Recôncavo Basin, and the other near Aracaju, capital of the state of Sergipe.[59]

The CNP had uncovered three new fields by the end of 1941: Candeias, Aratú, and Itaparica, all in the Recôncavo. The first two were of "commercial" size (a recoverable capacity of at least one million barrels, the minimum size from which it is economically feasible to extract oil); by that time it was evident that the Lobato field was not commercial.[60] The war, however, was now beginning to affect exploitation. The CNP found it increasingly difficult to secure equipment from the United States owing to the wartime shortage of ships.[61] While its budget was not large ($1.5 million annually for 1940–43), the shortage of parts meant the agency could not keep what little equipment it had in operation.[62]

Most of the petroleum Brazil consumed came from Aruba, in the Caribbean, and by mid-1941 Brazil was beginning to experience a fuel shortage, as fewer and fewer tankers could be diverted from war theatres. By July 1941 supplies of gasoline were down to two months and those of fuel oil to one month. The government thought seriously about rationing. The shortage was partly alleviated, however, by "gasoline" distilled from charcoal, known as *"gasogênio"*. The government had begun development of this fuel in 1938, and by 1941 felt it was sufficiently perfected to be sold as a substitute for petroleum-derived gasoline. The Gasogênio Law of July 15, 1941, compelled owners of fleets of more than ten trucks to use *gasogênio* in every tenth vehicle. The government agreed to sell machines to produce the substitute fuel at cost to public-service establishments and private concerns engaged in transport and agriculture. Some industries of this type installed machinery to burn wood for charcoal. As an economy measure the government had required the addition of anhydrous alcohol to gasoline for some years prior to 1938; the percentage of alcohol added increased as the war dragged on.[63]

In July 1941 Brazilians were asked to conserve fuel against an anticipated tanker shortage. By the end of August some interior cities were partially blacked out, owing to fuel shortages; industry suffered. Rio de Janeiro reduced its bus service and no gasoline was sold on Sundays and holidays. Rationing was in effect in Bahia and southern Brazil.[64] Then, in the same year, Standard Oil of Brazil sent directly to President Vargas another proposition couched in somewhat different terms. The new proposal complained that the current legislation inhibited the company in its desire to help in the search for Brazil's petroleum; for the company to do so would require "a constitutional amendment as well as certain modifications in the general laws of petroleum". The proposal recommended changes in legislation to guarantee a company the "reasonable possibility of profits proportionate to the risks" involved; Standard Oil of Brazil would then be prepared to *co-operate* with the government and private Brazilian companies, to the advantage of the national economy. The company proposed that it work through a *sociedade anônima* (equivalent to the English limited-liability company); the company would employ some Brazilians, but only in positions that had been specified at the time of the company's formation.

Horta Barbosa made the same objections as he had to the previous memorandum: both the Constitution and the Code of

Mines prohibited such activity by foreigners. He reminded Vargas that Brazil's "genuinely nationalistic" mining and oil policies were determined as much by considerations of national security and sovereignty as by economic factors. The Army General Staff, as before, agreed with Horta Barbosa, while the Cabinet was divided.[65] But General Eurico Gaspar Dutra, Minister of War and a strong nationalist, urged rejection and Vargas eventually tabled the proposal.[66]

Early in September 1942 Standard Oil of Brazil made yet a third proposal, similar to the second. Again citing defence factors, Dutra persuaded the Cabinet to turn it down.[67]

Commenting upon the three proposals in October 1942, Horta Barbosa said he saw double danger in the alienation of Brazilian oil resources to foreign concerns: not only were such companies less concerned with the particular needs of the Brazilian economy than with their own world-wide profits, but they were also to a dangerous extent agents of American foreign policy. To support the latter allegation he cited a 1926 recommendation of the United States Federal Oil Conservation Board, urging American oil companies to expand into Mexico and South America. The Board had said, "It is of paramount importance that our companies acquire these fields and develop them intensively, not only as a source of future supply, but of supply under the control of our citizens."[68]

These incidents do not portray Vargas as an ardent nationalist. Indeed, he preferred a "mixed" arrangement between government and private sectors to launch an oil industry (as had been created for the Volta Redonda steel mill). It was not until after his suicide in 1954 that Vargas became known as a champion of state monopoly. What these incidents do demonstrate, however, is the importance of army support to him. The majority of Brazilian officers at this time appear to have endorsed the corporatist nationalism Horta Barbosa championed, at least to a sufficient degree to balk at opening the oil industry to foreign exploitation. It is significant that Vargas would not press his own preference in the face of opposition from the army.

Brazilian transportation and industry continued to suffer as the war continued, because of national dependence upon oil imports. The occasional sinking of a tanker bound for Brazil always resulted in a fuel crisis; gasoline rationing increased and the number of vehicles on the roads declined. In May 1942, 3,000 private-car owners "voluntarily" gave up their gasoline ration to physicians and ambulance operators. After July 15, President Vargas banned all private cars, except those of high officials, and raised the proportion of anhydrous alcohol to be added to gasoline from 30 to 50 per cent. Factory after factory was forced to shut down for lack of fuel oil. In January 1944 the CNP ruled that no new licences would be issued for trucks or cars, and the Ministry of War restricted the use of army vehicles.[69]

Meanwhile the growing scarcity of parts had forced the CNP to reduce its exploration effort, and by 1942 it could explore only in the Recôncavo.[70] The private Brazilian companies exploring in various states had no success; their efforts were described as "scattered".[71] By the end of 1943 national production had reached only 300 b.p.d., approximately 1 per cent of consumption.[72] To many it must have seemed incredible that a country so desperately short of fuel should be vigorously refusing expert foreign aid that might accelerate the discovery of domestic sources of oil. When Horta Barbosa suddenly resigned from the CNP on August 16, 1943, it seemed to indicate that Getúlio Vargas might be contemplating a new petroleum policy.[73]

3

LIBERALISM THWARTED, 1943–1950

The period 1943-50 saw a sharp revision in the policy of economic nationalism that had been developing in the Brazilian petroleum industry up to that time. The new CNP president, Army Colonel João Carlos Barreto, who had replaced Horta Barbosa in 1943, was more concerned with rapid oil development than with the maintenance of nationalistic controls over the fledgling Brazilian industry. Moreover, he was openly receptive to the idea of foreign participation—even by "the trusts" — in finding and exploiting Brazil's reputed sheets of oil. Accordingly, the postwar era saw a new government attempt to open Brazil's oil reserves to a measure of foreign exploration and development.

Earlier, however, in a bid to stimulate exploration, the Minister of Agriculture in 1943 proposed a new mining code, by which the landowner would be guaranteed preference in subsoil exploration and development and a share in profits if exploitation were by a third party. President Vargas approved the idea, but the committee appointed to draft the legislation could not reach agreement and the proposal was dropped.[1] As one commentator remarked, there had been no activity under such a regime in the 1920s and there was no reason to expect a change in 1943. "It would," he said, "be like trying to revive a cadaver."[2]

Vargas made another attempt to resurrect oil exploration in 1944, promulgating a decree permitting foreigners to subscribe up to fifty per cent of the shares in mining companies and to elect one less than half of the directors. But the decree ran counter to Article 143, Subsection 1, of the 1937 Constitution, which prohibited foreigners from engaging in mining activities. Vargas's measure was thus never implemented. That the decree was even formulated, however, indicates that Vargas and Barreto were of one mind on oil development.³

Barreto also tried to prod the exploration effort of the CNP by bringing in Everett L. DeGolyer and Lewis W. MacNaughton, of the American consulting firm of the same name, to recommend priorities. Barreto also announced that the CNP would hire foreign technicians to assist both in policy-making and in training Brazilians, and that more Brazilians would be sent out of the country to keep abreast of technical developments in the industry. Barreto was able to use foreign firms to give some coherence to the CNP exploration effort: DeGolyer and MacNaughton supervised technical work, the Drilling and Exploration Company organized exploration (using Brazilian technicians, principally), and the United Geophysical Company continued to direct geophysical exploration (as it had under Horta Barbosa). Following DeGolyer and MacNaughton's recommendations, Barreto concentrated the CNP's meagre resources in the Recôncavo. Unfortunately, however, orders placed in the United States for more sophisticated equipment simply were not filled, probably due to war priorities there.⁴

Meanwhile, production declined relative to consumption. By the end of 1945 the average was 217 b.p.d., or 0.7 per cent of Brazilian consumption.⁵ Although the CNP appropriation rose to $2.5 million for each of the years 1944 and 1945, the wartime scarcity of parts kept much equipment idle.⁶ But another factor retarding production was the quality of oil found in Bahia: it contained a high percentage of paraffin (over 20 per cent in crude from the Lobato-Joanes field), which constantly clogged pipes and storage tanks, necessitating expensive steaming and cleaning at frequent intervals. The oil was so heavy that it required considerable cracking to yield significant quantities of gasoline; the cheaper and simpler topping* process yielded little gasoline and left the majority of the crude in the form of a highly

---

* Cracking: the process of breaking down by heat and pressure the heavy, complex hydrocarbons of petroleum into the lighter and simpler hydrocarbons of gasoline.
  Topping: distillation of crude oil to separate it into its fractions.

viscous residue that was of little use, even as fuel oil. It was hardly worth increasing production since the product was so difficult to handle. The few tiny distilleries that existed were not, of course, equipped to handle Brazilian oil, and not until 1943 and 1944 did the CNP build small refineries (150 b.p.d.) at its Aratu and Candeias fields, to supply fuel for rigs and trucks operating there.[7]

On May 6, 1945, the CNP issued a policy statement admitting that the great amounts of capital needed for the petroleum industry "cannot yet be found . . . in a strictly nationalistic orbit." Since industrialization required an adequate national supply of petroleum, the statement continued, and since foreign capital was available to help develop national oil reserves, constitutional and legal barriers to such assistance should be modified. It should be possible to balance internal and external interests, always keeping in mind national security. Accordingly, the CNP recommended that foreign capital be attracted, but that the nation retain ownership of subsoil wealth; that exploration and exploitation continue to be through federal authorization and by Brazilian citizens only; that no monopoly of the industry be allowed; and that preference in all phases of the industry be given to Brazilian capital.[8]

Later that year the CNP sought to interest Brazilian capital in establishing oil refineries, and encouraged applicants to invest up to 25 per cent of refinery profits in exploration. By July 1946 four groups had agreed to construct refineries (two in the region of Rio de Janeiro and two near São Paulo).[9] The CNP estimated that a refinery of 10,000 b.p.d. would cost up to Cr$150 million ($7.5 million), which would yield an estimated gross profit of Cr$56 million ($2.8 million). To reserve 25 per cent of profit, less certain obligatory deductions, would yield Cr$12 million or enough to explore effectively 20,000 hectares (49,400 acres), the CNP estimated.[10]

Meanwhile, the CNP had brought in a well in the Bahia Recôncavo producing 1,800 b.p.d. Its morale thus boosted, the agency planned to spend the expected surplus income by stepping up exploration in the Recôncavo and building a refinery near Salvador, Bahia, with a capacity of 2,500 b.p.d. to process local crude. To finance the project, the CNP formed a mixed company with a capital of Cr$50 million ($2.5 million), half of which national capitalists could subscribe.[11] Even Sílvio Fróes Abreu and Glycon de Paiva, two of the geologists who had braved the scepticism of their colleagues to urge the Lobato exploration ten years before, agreed that foreign capital, participating under strict government control with mixed national companies, was necessary for the growth of the Brazilian petroleum industry. What they felt was still especially needed was incentive for private assistance in such costly ventures as exploration.

Fróes Abreu, in particular, charged that the insularity of Brazilian technicians had acted as a brake on national oil exploration and development, and lack of contact with foreign geologists meant that Brazilian methods were behind the times. Other commentators assailed Brazilian investors as not sufficiently risk-conscious, and asserted that foreign capital was necessary for the creation of a viable national oil industry. The disinclination to take risks in investment, the charge concluded, emanated from a "civil-service mentality", by which Brazilians sought a secure, stable sinecure, which carried little responsibility. Not the stuff of great oil barons![12]

The 1946 election, which brought General Eurico Dutra to the Brazilian presidency, signalled a relaxation from the privations and strictures of the Estado Nôvo. Elected to continue Vargas's policies, Dutra proved bland and non-political. Appealing for a return to "tranquility," he initially tried to restore *laissez-faire* liberalism as a remedy for war-induced inflation. Brazil would turn away from state intervention, identified with both the war and the previous authoritarian regime, and welcome imports of all kinds from abroad.[13]

Dutra's non-political stance apparently extended to petroleum, since he did not oppose the clauses which relaxed existing strictures on oil development, even though three years earlier he had pressured Vargas into rejecting a proposal from Standard Oil. The new posture, which the army obviously endorsed, also reflected the postwar eclipse of fascism in the Western Hemisphere; Horta Barbosa's corporatism had gone out of fashion. So it is no surprise that the 1946 Constitution, framed in the aftermath of wartime oil rationing, opened the way to broader participation in the national petroleum industry. The constituent Assembly that formulated the document was deeply concerned to "impede the perversion of the concept of nationalism, the use and abuse of which had produced in the world the dramatic results we had been forced to witness." Its only restriction on petroleum development was that such activity should be carried out by Brazilians or "companies organized in the country"; no stipulation was made about the nationality of shareholders in such companies.[14]

Since there was a political climate apparently receptive to private and foreign enterprise, coupled with incentives from the CNP toward the same groups to establish refineries and engage in exploration, a rush of proposals might have been expected. The rush, however, did not materialize, and for very good reasons. First, the CNP set unduly harsh terms for awarding refining permits, terms which forced the withdrawal of two of the four bids made the year before. The CNP required from applicants, within ninety days of January 18, 1946, a deposit of Cr$500 ($25) per barrel of projected refining capacity, the submission of a firm contract for a five-year supply of crude oil, and plans so complete that construction could start within thirty days of approval and be finished within two years. Moreover, 9 per cent of the CIF price of the crude oil imported was to be contributed to exploration (roughly 50 per cent, in the estimation of the CNP, of projected net profit); the company could remit this to the CNP, rather than carrying out exploration itself, and the CNP would return to the company an appropriate percentage of any oil discovered. Harsh terms indeed, which were designed to scare off all but the most serious contenders. The CNP clearly wanted to forestall the entry of the fly-by-night private companies that had characterized exploration until the formation of the CNP. By 1946 Brazil had only four refineries, the largest of which had a capacity of only 1,000 b.p.d. Three were located in Rio Grande do Sul: two of them were operated by Uruguayan and Argentine interests, and the third, Brazilian-operated, bought crude from the other two. The fourth, in São Paulo, was operated by the Matarazzo industrial network for its own consumption. The CNP, however, wanted more viable and significant refineries with capacities of 10,000 b.p.d.

A second reason for the lack of an "oil rush" in Brazil was that the United States lost interest in Latin America after the Second World War, when it turned to reconstruction in Europe. Loans through the Export-Import Bank were no longer available for such ventures as building refineries in Brazil, and there were virtually no other creditors to turn to. The international oil companies, "the trusts" of recent vilification, were not particularly interested in Brazil with its relatively unattractive geology, small market, restrictive legislation, and history of hostility to foreign oil companies.

It was not surprising that only two companies survived to register bids for refineries. One was headed by Alberto Soares Sampaio, who had made his money in railroads. He had strong financial backing from São Paulo, and, initially, an agreement from Standard Oil of New Jersey to supply equipment, crude oil, and 49 per cent of the capital required. Soares Sampaio proposed building an 8,000 b.p.d. refinery at Rio de Janeiro. Unfortunately, after the 1946 terms were announced, Standard withdrew, and the group—Refinaria e Exploração de Petróleo União—allowed the CNP to persuade it to relocate at Capuava, São Paulo, in return for an increase in capacity to 20,000 b.p.d. The second company, headed by Drault Ernany de Melo Silva, a Rio banker, had exclusively Brazilian financing. It was able to secure a contract for a crude-oil supply only after the U.S. Ambassador, Adolph A. Berle, pressured Standard Oil of California into the deal. This group also wanted to build at Rio, and held a lease on sufficient land at Manguinhos, near the city, an ideal site for a refinery of 10,000 b.p.d.[15]

Thus, by 1947 the two private companies had cleared several serious hurdles to arrive at the starting point. Yet several more remained, which would further delay construction for many years. Barreto, perhaps chagrined that his policy had not been conspicuously successful, decided in 1947 to try to lure private capital into the oil industry by a partnership with public capital. He proposed a mixed company to build a refinery to process Bahian crude oil at Mataripe, near Salvador. What made the project feasible was the discovery, early in the year, of the Dom João field, which had oil at a very shallow depth and of a light quality, with a lower percentage of paraffin than had been found elsewhere in the Recôncavo.

By the end of 1947 the CNP estimated that its maximum potential production was 9907 b.p.d.; even when two-thirds of this was taken away (a reduction due to paraffin content: lines and tanks had periodically to be emptied, steamed, and cleaned, or they would quickly become clogged with wax), potential production still appeared sufficient to justify a refinery of 2,500 b.p.d. capacity.

The CNP then contracted with the W. Kellogg Company to build a refinery of that capacity in November 1947 at Mataripe (close to the Candeias field, from which almost 90 per cent of production came). The estimated cost was $2 million, all in hard currency, since no materials would come from Brazil.[16] Unfortunately, no private subscribers came forward, and Barreto's planned Refinaria Nacional de Petróleo S. A. languished for two years. Despite the relatively disappointing results from Barreto's new departures in oil development, however, it was clear that many Brazilians, particularly in the higher ranks of the military close to President Dutra, endorsed Barreto's approach to the problem. In their view, Horta Barbosa's rigidly nationalistic stance was as anachronistic as the Estado Nôvo in those democratic days; they promoted a more liberal approach, while still insisting on state regulation and control.

On February 14, 1947, the CNP convened a Comissão de Anteprojeto da Legislacão do Petróleo (Committee for the Draft of Petroleum Legislation) to review existing oil laws, in the spirit of the constitution and "practicality". Former Minister of Agriculture Odilon Braga, who had convened the Petroleum Inquiry Commission in 1936, headed this committee. It also included Professor Ruy de Lima e Silva, Avelino Ignácio de Oliveira, Glycon de Paiva, Colonel Arthur Levy, and General Antônio José Alves de Souza, all of whom had been intimately associated with Brazilian oil for years.

At the same time Dutra proposed to the Chamber of Deputies that foreign participation in oil refining be permitted, up to a limit of 40 per cent of the share capital. He cited three reasons: his concern to stimulate the national economy and provide for national defence, economy of foreign exchange (greater than Cr$400 million, had Brazil then been refining all the petroleum she imported), and "the necessity of avoiding new crises in transportation, produced, in part, by lack of supply of oils and derivatives."[17] The postwar government clearly endorsed Barreto's recipe for the development of an oil industry, and was as clearly anxious to abandon the kind of economic nationalism associated with Horta Barbosa.

Odilon Braga told the Engineering Club on May 7, 1947, that the task of his committee was to frame laws which would attract foreign capital to the Brazilian oil industry, while balancing the extremes of state monopoly and free exploitation by concession. "Firm, clear and adequate official control" over all phases of the industry would be needed, he added, so that oilfields would remain the inalienable property of the state. He recommended that no transfer of property be involved in exploration and development, that the companies be organized in Brazil and subject to its laws, and that supply of the internal market be under Brazilian control. He believed that foreign companies would be attracted by opportunities to explore large areas with assured exploitation of fields discovered, a high profit margin, the certainty of a quick return on capital, and guarantees against any encroachment upon their rights. Even Horta Barbosa, concluded Odilon Braga, would be satisfied with the degree of official control he had described.[18]

As further proof of a trend toward liberalization, General Juarez Távora then began a campaign to open petroleum reserves to private development, national and foreign. Távora had been a *tenente* in the 1920s, a leader of the 1930 Revolution, and Minister of Agriculture from 1933 to 1935. He had then been isolated politically during the Estado Nôvo, which may have contributed to his feeling after the war that such rigid nationalism was out of date. He said that he had been a strong nationalist until the effect of the war upon Brazil's petroleum supplies had changed his mind. Given the decisive role oil had played in the Second World War, said Távora, no country could hope to defend itself until it had resolved its petroleum problem. He emphasized how much Brazil owed the Allies for the defence of Christian values in the war. The Atlantic Charter and Resolution XI of the 1945 Chapultepec Conference had called for economic co-operation among the Americas and equality of access to raw materials. In the midst of such a spirit, asserted Távora, Brazil should not adopt a policy of "pseudo-nationalism", which had been the ruin of many other nations. Brazil should allow foreign capital to participate in its oil industry, and Great Britain and the United States (the principal Allies) would be the countries from which most of such assistance would come.[19]

The Brazilian government, Távora suggested, had repeatedly revealed its lack of capacity for management on a national scale. Given the size of the area to be explored and the urgency of requirements, Brazil needed foreign capital, equipment, and technicians. He urged that exploration and development be open by concession to companies which were national, wholly foreign, or a mixture of both, but that transportation and refining be reserved for national companies, or mixed enterprises at least sixty per cent Brazilian-owned. Export of crude oil or refined products would be permitted only after domestic needs had been satisfied. In case of expropriation, companies would be assured of indemnification to the amount of invested capital not yet amortized. Companies would be required to turn over to the state without indemnification one-half of any oil-bearing areas encountered, to become national reserves; exploitation of the other half should not prejudice the reserve area.

Távora's statements prompted a rash of endorsements from public figures. In some instances the CNP itself came under fire as an inhibiting factor in the growth of a Brazilian oil industry, and in particular of exploration. One commentator charged that under Horta Barbosa the CNP had been a dictatorship, absolutely hostile to private participation in the industry while at the same time incapable of mounting a significant effort of its own.[20] It was hardly to be expected that such sweeping changes in the policy of economic nationalism would go unchallenged, and in July, Horta Barbosa embarked on a series of lectures to promote a state monopoly for petroleum.[21] His theme, strongly corporatist, was, "Petroleum belongs to the nation, which must distribute it, equally, for its children."[22]

For Horta Barbosa, history revealed that "where state monopoly existed, as in Mexico since 1938, and above all in Argentina since the discovery of the first well in 1906, the entire nation — not isolated groups — was the beneficiary." Where private monopoly prevailed, he continued, as it did in Venezuela and Paraguay, "only the exploiting companies and not the people enriched themselves from high production", and the nations continued in a "primitive state of economic and political dependence". National reserves could not survive under the concession system, since there would be no way of preventing a concessionaire from depleting the reserve-half of a field from wells on his own half. Finally, foreign companies were more interested in profit than in the good of an individual nation's economy.

As a basic step, Horta Barbosa recommended the creation of a national refining industry, "with or without the discovery of oil fields". National refineries could import crude "so long as the market was free". Exploration could be financed with profits from refining, and future national oil discoveries could be processed by the existing refineries. "The trusts" should have no part, Horta Barbosa said, in any phase of a nation's oil industry; once they had secured a foothold they would irresistibly take over the entire operation. To talk of a mixed company, for Horta Barbosa, was treason: Brazil could not hope that its citizens who were included in an enterprise with "the trusts" would safeguard the collective welfare of Brazil, and resist the attraction of dividends and other blandishments. Soon the state and "the trusts" would be at war and it was "easy to see who would be the victor." He quoted U.S. President Woodrow Wilson, who had said, "The value of a nation depends on the quantity of petroleum it possesses." Any government that alienated its petroleum was contributing to a decline in the value of its own country to the benefit of another.[23]

The battle lines were drawn. Juarez Távora argued for liberalization and Horta Barbosa for carrying economic nationalism to its logical conclusion: state monopoly. The fact that both men were high-ranking army officers shows how deeply the military was divided on oil. Because of its pivotal position in politics and because petroleum was vital to national defence, the military establishment was cast in a key role in the debate.

The first legislative proposal for a petroleum law was presented in the Chamber of Deputies, not by Odilon Braga's committee but by the Communist bloc. On June 25, 1947, Deputy Carlos Marighela proposed the creation of the National Petroleum Institute (INP), replacing the CNP. Unlike the CNP, which existed solely on budget allotments, the INP would have "absolute monopoly over all phases of the industry", and its main source of revenue would be from the distribution and sale of petroleum products. It could form mixed companies (which it would control), with government and private Brazilian capital only, to operate the oil industry.[24]

Apparently not satisfied with this plan, the Communist bloc drew up a second proposal which, by contrast, permitted *only* private Brazilian capital to invest in national petroleum companies. The policy switch was to gain popular support. The party referred to its new plan as the "Monteiro Lobato Thesis", in an effort to capitalize on his enduring popularity as an oil promoter.[25] But soon the Communists dropped both proposals to await the legislation being drawn up by Odilon Braga's committee.[26]

The *Jornal de Debates*, a nationalistic weekly, turned its attention to the emerging controversy in mid-1947. In a long article on August 22, Fernando Luiz Lôbo Carneiro, a civil engineer and former employee of the CNP, argued that since Brazil had already discovered petroleum, she did not need the vast financial resources of "the trusts" but would be better off with her own refineries and tankers, from the profits of which she could finance an exploration program.[27]

Little is known about Lôbo Carneiro. Undoubtedly he was a Communist, and the Communist Party of Brazil had switched to a cold-war policy by this time (late 1947). It was striving for mass support and seizing opportunities — such as attacking the proposed liberal oil policy — to discredit the government.[28]

Next the *Jornal de Debates* reprinted a speech, given under the auspices of the Brazilian Socialist Party (PSB), by journalist Rafael Corrêa de Oliveira. He was less restrained than Lôbo Carneiro. First the great international companies had asserted that Brazil had no petroleum, said Oliveira; now they were saying not only that Brazil did have petroleum but that the fate of Christian civilization depended upon its immediate exploitation. Their argument, said Oliveira, was that the United States and Great Britain were faced with war with Russia over the oil of Iran, and Brazil could not refuse the international collaboration offered for the rapid development of its vast petroleum reserves. But assistance, Oliveira noted, was offered in the form of Standard Oil.[29]

In October 1947 the *Jornal de Debates* published a speech Lôbo Carneiro had made to the National Student Union, which expanded his earlier attack. After reminding his audience that the great international oil companies had previously been free to explore in Brazilian territory but had not done so, he quoted the U. S. Secretary of Defense, James Forrestal, to the effect that the oil fields of Saudi Arabia were much more important to the United States than those of the Western Hemisphere. "The trusts," charged Lôbo Carneiro, had no intention of developing Brazilian oil; they merely wanted "to keep Brazil from resolving her petroleum problem on her own," so as to retain the lucrative Brazilian market.

"It is therefore urgent," he declared, "that public opinion be mobilized, not only to impede the surrender of our petroleum... to the foreign trusts, but also, and this is no less important, to force a rapid solution to the problem with our own resources." He went on:

> ... it is necessary that there be created in Brazil a genuine mystique around the petroleum industry, as in Argentina. I am convinced that if the state-monopoly solution should be imposed upon the government by a vast and profound movement of public opinion, it could not miscarry, since the entire Brazilian populace would follow its course with affection, and ensure that its aims were realized.
>
> For this campaign, the idealism, the enthusiasm and the optimism peculiar to the young are indispensable.
>
> The petroleum campaign should give new spirit to Brazil. The Brazilian people are not inferior to the Argentine, the Uruguayan, the Chilean. We ought to repel vehemently the insults of incapability, indolence or corruption hurled against us. We will know how to triumph in the battle of petroleum.[30]

Soon after this speech, the Centro de Estudos e Defesa do Petróleo e da Economia Nacional (Study and Defence Centre of Petroleum and the National Economy, hereafter referred to as the Petroleum Centre) was organized to force the adoption of a state monopoly for the petroleum industry. In all probability, Lôbo Carneiro played an important role in its formation. Members included many of the strongest nationalists from the military and public services.

At the same time, street demonstrations for state monopoly, based on the slogan "*O petróleo é nosso*" (the oil is ours), began among university students (first in Rio). Both the Petroleum Centre and the "O petróleo é nosso" campaign claimed inspiration from Horta Barbosa's addresses.

By late 1947 it was becoming obvious that "O petróleo é nosso" had substantial appeal for much of the Brazilian populace, since a campaign in its support began almost spontaneously and quickly assumed national proportions. What is even more striking is that the campaign was not just a "flash in the pan", serving merely to bring down an attempt to liberalize oil laws (which it did); it persisted until 1964 in the political arena and goes on even today, although outside politics. Thus, we must examine to whom such a slogan appealed and why they found it appealing.[31]

The most significant group of supporters for "O petróleo é nosso", in importance if not in numbers, was the military. The majority of officers strongly backed a state monopoly over oil for security reasons and out of a corporatist-inspired urge to maintain national sovereignty over natural resources. A minority of officers, led by Juarez Távora, plumped for liberalization; they tended to be closely associated with the Escola Superior da Guerra (Higher War College — ESG) and to be veterans of the Brazilian Expeditionary Force which fought in Italy during the Second World War. From these experiences they had acquired a strong antipathy to fascism and excessive state control, and a belief in continentalism and in the willingness of the United States to help their "good neighbours to the south".[32]

A second group promoting "O petróleo é nosso" comprised student organizations at the university level. For them the nationalism of the slogan was extremely appealing. The University Reform Movement had stirred the social conscience of two generations of Brazilian students; the social theory of reform taught that national resources should be at the service of the state, "for its children" (as Horta Barbosa had said). Furthermore, the industrialization begun by Vargas during the Estado Nôvo meant a shift in the economic power structure and more job opportunities for graduates; to revert to a "colonial" economic policy (i.e., foreign control of the national oil industry) cast a shadow over the future of many students.³³ And such old-style nationalists as ex-President Artur Bernardes viewed a liberal oil policy as an admission that his countrymen were not capable of developing their own resources, an interpretation that stirred those who hoped to play a role in that development.³⁴ Lastly, Lôbo Carneiro's speeches seemed to call upon the students to lead a willing populace against a treacherous government; this message could not fail to appeal to those students who nursed political ambitions (as many of them did).

A third group of supporters was made up of intellectual economic nationalists who wrote in such newspapers and journals as *Última Hora, Diário de Notícias, Jornal de Debates*, and later *Cadernos do Nosso Tempo* and *O Semanário*, and who were instrumental in keeping the campaign alive and in the public eye for so many years.

None of the substantial groups mentioned above, however, was monolithic in its view of economic nationalism; each embraced several more-or-less-radical philosophies of economic development. Basically, these philosophies can be reduced to two, which Thomas E. Skidmore has called "developmentalist nationalism" and "radical nationalism". The former was most articulately defined and promoted by the United Nations Economic Commission for Latin America; in Brazil, its intellectual centre eventually came to be (in the mid-1950s) the Higher Institute for Brazilian Studies (ISEB). In the late 1940s this philosophy had no outstanding mouthpiece but took the form of a moderately nationalistic approach to development, favouring mixed companies under state direction and control. It was and is widely endorsed in Brazil. "Radical nationalism", on the other hand, was less an economic formula than a political polemic: it characterized the current economic structure as "exploitative" and demanded radical change. Radical nationalists claimed that a conspiracy existed — comprising the developed nations and those Brazilians engaged in the export-import trade and those employed by foreign firms—to keep Brazil in a colonial position, inhibiting any attempts at industrialization. The only good economic enterprise, according to such thinking, was under total state control.³⁵

But the campaign also enjoyed the support of a large proportion of the Brazilian population, and appears to have cut across class lines. The reasons for this are perhaps less tangible than those associated with the above-mentioned groups, but no less important. Probably most important was the sense of political release after the Vargas dictatorship: not only was there a residue of nationalism among the masses (Vargas had, of course, used nationalism to justify his rule), but the appearance of democracy suddenly added to this legacy a sense of participation in the national destiny. The oil issue was a superb focus—the first after the removal of Vargas—for this new awareness of national duty.[36]

Then, too, many Brazilians believed that their country had to modernize quickly to assume its rightful place as a leader of the hemisphere and a great power. Vargas had begun industrialization—the surest path to modernity—and the task now was to step up the pace. Brazilian private enterprise had demonstrated that it was either unable or unwilling to lead the drive, however, and the experience of other countries had shown that foreign developers were interested only in leeching profits. The solution was state control and initiative, with or without the co-operation of Brazilian private capital.[37] The type of oil development that Juarez Távora was proposing, therefore, was, in the above context, a regression to a colonial economy: Brazil would inevitably lose control of a vital source of energy.

Another factor contributing to widespread opposition to liberalization was the discovery of oil in 1939 by the state oil agency, the CNP, after its formation had signalled the adoption of a nationalistic oil policy. Orators in the late 1940s alleged that Brazilians, in the service of the state, had discovered oil after "the trusts" had declared there was none. Why then hand it back to the same parasitic groups? Brazilians, of course, feared "the trusts", as did many other Latin Americans at the time. As Peter Odell notes, oil has played a dominant role in hemispheric development, and the large international companies have been involved in Latin America from the beginnings of the industry. Relations with governments have never been easy, and companies have frequently been castigated—with some justification—as "agents of economic imperialism". However, coupled with Brazilians' understandable fear of "the trusts" was an intriguing resentment toward the same groups for *not* having tried to exploit Brazil's reputedly vast oil reserves.[38]

Finally, many Brazilians endorsed simple xenophobia in their support of "O petróleo é nosso". Put simply, they said that the Brazilian was the equal of any foreigner, and just as capable of developing an oil industry as anyone else.[39] Yet the draft oil law, the Petroleum Statute (Estatuto do Petróleo), that Odilon Braga presented to the Chamber of Deputies on November 11, 1947 (see Appendix II), ignored the nationalistic campaign. Braga said the committee wholeheartedly agreed with Juarez Távora's emphasis on continental development and on Brazil's need for foreign assistance to develop her petroleum industry. Braga criticized Horta Barbosa for thinking that Brazil could finance an exploration program with the profits from refining. Horta Barbosa's promised price reductions would also reduce profits, he said, and the higher operating costs that usually followed nationalization would further reduce earnings.

The key to management of a national petroleum industry, said Braga, was to recognize where the state's faults produced less inconvenience than the cupidity of private companies. For Braga, the state had more continuity of direction, but more rigidity and less response to opportunity than private companies. The state was less concerned with costs, and its agents exhibited less zeal and less concern for the financial and economic repercussions of their operations. Rigid and inflexible state accounting methods often stifled programs or smothered them altogether. But, he added, a combination of the two systems had permitted England, France, and Holland to achieve all the objectives of the great American "trusts" without the "excesses" of the American combines, without the sacrifice of many worthwhile enterprises, and with the control and profitable participation of the state.

The committee had rejected legislative proposals solicited from Standard Oil of Brazil and from Shell–Mex Brazil, the proposed Peruvian petroleum law, and the Venezuelan oil law of March 13, 1943—all markedly similar. All four postulated that the state owned all oilfields, but in practice this meant that the state retained only legal domain and the concessionaire exercised useful domain. Each proposal permitted the holding of great blocs of concessions. Finally, in all proposals the concessionaire exercised "sovereignty" over his concession. This left the state with only those functions "inherent" to it, principally the collection of revenue.

Braga said that the committee had taken the 1934 petroleum law of Great Britain as a model for the Petroleum Statute. The British law gave the state a positive role. The Crown owned all petroleum and enjoyed the exclusive right of exploration and exploitation, but concessions for "exploration, drilling and exploitation" were granted to competent applicants and the concessionaire became the owner of oil extracted under such conditions. Braga warned that Brazil had to act quickly. Foreign interest in its oil possibilities would probably decline within the next ten years, with the exploitation of the "tremendous" Middle East fields. The great oil companies, Braga contended, wanted to extract as much oil as possible from the countries bordering upon the Soviet Union. He pointed to Colombia's known, but dormant, fields as proof of a shift in their interests.[40]

Indeed, in retrospect it is clear that it was too late for Brazil already. Braga, Távora, and the supporters of the bill conceived of oil development in terms of wartime security and cooperation. But by the end of 1947 the United States was, to all intents and purposes, completely involved in European recovery, and the oil companies were disinclined to be generous toward Latin America, when there were greener (blacker?) fields elsewhere in the world. The Statute was doomed to emerge stillborn. The National Security Council of the Chamber of Deputies agreed that the proposals complied with the "fundamental interests of the economy and defence of the nation" and would attract foreign capital, while giving preference to national capital and ensuring that the national market would be adequately supplied.[41] The Investments Committee, however, after having endorsed Braga's earlier outline of the Statute, changed its mind and recommended stricter government control over the industry, with gradual nationalization of all its phases.[42]

The Chamber of Deputies asked two American oil experts, Herbert Hoover, Jr., and Arthur A. Curtice, for their comments. They doubted that a foreign company would invest in exploration when, after completing a program, there would be no guarantee that it could exploit the area. They also doubted that private Brazilian capital would be interested, under the circumstances. Furthermore, since the bill ruled that concessions could be rescinded by the state at any time with indemnification based upon historic costs only, the risk would frighten off any company, national or foreign. Hoover and Curtice submitted that there were many countries with better oil potential and more amenable legislative climates than Brazil. Since Brazil was just starting to develop a petroleum industry, with so many of the important factors governing growth unknown, Hoover and Curtice urged that it set up the most liberal terms possible to attract foreign capital.[43]

Juarez Távora also criticized the bill. He claimed that in 1946 "small companies" and individuals had drilled fully three-quarters of the approximately 30,000 oil wells sunk in the United States, the land of "trusts" and giant companies; over 80 per cent of the strictly wildcat drilling in that year had been done by small operators. The role of the big companies was less important than was generally believed, said Távora, and he begged Brazilians to think about this before eliminating private activity from a national petroleum industry (which, he submitted, the Petroleum Statute would do). Távora claimed that at heart he agreed with Horta Barbosa, favouring state monopoly. But he believed — especially as a member of the General Staff — that such a policy would retard solving the problem of national supply. No one could guarantee that Brazil would not be confronted with a war situation in the near future, unable to defend itself for lack of oil.[44]

The appearance of the bill touched off an emotional debate in the press. Of the principal newspapers in Rio de Janeiro, *O Jornal* favoured state monopoly, providing space for articles on the theme by Deputies Juracy Magalhães (ex-*tenente* and *Interventor* of Bahia) and Artur Bernardes (ex-President of Brazil and a xenophobic nationalist). *Correio da Manhã*, a staunchly "liberal" paper, supported the Petroleum Statute and Juarez Távora, while the fiercely nationalistic *Diário de Notícias* reported the activities of the Petroleum Centre and carried the articles of such proponents of state monopoly as Rafael Corrêa de Oliveira.[45]

Individual journalists reflected the diversity of opinion. Perhaps the most widely read proponent of the "Távora thesis" was Jorge Abdalla Chamma, who called for the controlled use of massive amounts of foreign capital so that Brazil could get its petroleum out of the ground.[46] Matos Pimenta, a frequent contributor to the radical-nationalist tabloid, *Jornal de Debates*, as well as daily newspapers, said Brazil "should make every effort and concession to entice private foreign capital into the country." Yet he could not sanction handing over Brazilian sources of energy to private companies, national or foreign. Pimenta cited Pope Pius XI who, in the encyclical *Quadragesimo Anno* (1931), had declared that such reserves should be exploited by the state; moreover, Matos Pimenta said he feared the power of the great oil companies, "which control even the United States Government."[47]

Finally, articles by Juracy Magalhães represented radical nationalism: his constantly reiterated theme was that the bill was the work of Standard Oil.[48] Hélio Lacerda took Magalhães' argument to extremes, calling the Petroleum Statute "a monstrous crime against the country", which tried to hide its treachery in woolly phrasing. He urged people to organize petroleum-defence committees to fight the bill. The Americans, Lacerda wrote, considered Brazil one of the world's richest countries in oil. Oil was seeping out of the ground all over the country and Juarez Távora was wrong to say that Brazil needed vast sums to extract it. It was abundant and easily accessible, and, as Horta Barbosa had said, the people's money was sufficient to bring it out of the ground.[49]

The Clube Militar in Rio de Janeiro, which, because of the persistent presence of the military in Brazilian politics, is and has been an important forum for the discussion of national issues, was at the centre of the debate over oil policy. Both Horta Barbosa and Juarez Távora launched their campaigns there, and the club remained deeply involved in the controversy for many years. It sponsored addresses by military officers on the question, and its review, the *Revista do Clube Militar*, contributed articles, comments, and news on the subject. In 1948 direction of the Clube Militar was in the hands of radical nationalists, General Salvador César Obino as president and General Estévão

Leitão de Carvalho (one of the leading figures in the Petroleum Centre) as vice-president, both of whom staunchly supported Horta Barbosa. By mid-1948 it was clear that the review had adopted the same view. One 1948 article was entitled "The Petroleum Problem in Brazil: Against the Imperialistic Offensives of the Trusts", and pointed to the "disastrous examples" of their activities in many countries. General José Pessoa, the author and a former president of the club, opposed participation in any form by these companies in the Brazilian oil industry.[50] Soon afterward, "Captain X" first appeared in the review (his identity was never publicly revealed). The "captain" followed a policy of rigid and extreme nationalism and never tired of citing the arguments of Horta Barbosa. His theme was that once "the trusts" were allowed in, they would ultimately control the industry, no matter how small their share of the total capital.[51]

Some of the technical journals espoused Juarez Távora's thesis. *Mineração e Metalurgia (Mining and Metallurgy)*, for example, charged that the slogan "O petróleo é nosso" was Communistic, inspired by Moscow. The oil, asserted the journal, is not "ours" until it has been discovered and exploited.[52]

But not all professional men, or their periodicals, subscribed to the "Távora thesis" without question. The *Revista do Clube de Engenharia (Engineering Club Review)* usually endorsed the participation of private capital in the oil industry, but an occasional contributor spoke out against it. R. Descartes de Garcia Paula, for example, agreed that from a strictly technical standpoint "the trusts" would be the best agents to explore and exploit oil reserves, but he could not banish his fear that they might put their own monetary interests above those of Brazil. He also feared even the limited entry of foreign capital into the oil industry as a risk to security, and so was forced to support state monopoly.[53]

Hermes Lima, a socialist advocate of state monopoly, opened the debate over the Petroleum Statute bill in the Chamber of Deputies on March 19, 1948. He asked its supporters whether, if Odilon Braga had been correct in saying that "the trusts" were more interested in the Middle East, they really expected foreign capital to invest in Brazilian oil. Brazilians, he said, would not be willing to place their petroleum industry at the service of the

United States in the event of war, and he warned that country that if "the trusts" participated in the development of a national petroleum industry, Brazil would be swept by a serious wave of anti-Americanism.[54]

Carlos Lacerda, just beginning his career as a journalist (he later became one of Getúlio Vargas's most bitter opponents and an effective, if highly controversial, governor of Guanabara state), ridiculed Hermes Lima. Writing in the generally liberal newspaper *Correio da Manhã*, he scoffed at Lima's implication that Brazil was too weak to confront "the trusts" under the Petroleum Statute. Such a nation, he charged, could not hope to undertake the far more onerous task of creating a national petroleum industry.[55]

Debate continued throughout 1948. Euzébio Rocha of São Paulo and ex-President Artur Bernardes of Minas Gerais—both men unrelenting economic nationalists—emerged as the principal spokesmen in the Chamber of Deputies of Horta Barbosa's state-monopoly thesis. Rocha based his attack on the loss of sovereignty implied in granting concessions to a foreign oil company: the worth of Brazil would decline in proportion to each drop of oil extracted under such a system.[56] Bernardes, as staunch a nationalist as he had been thirty years before, attacked the bill because it cast doubt on the abilities of his countrymen: "We are trying to discover whether or not the Brazilian is capable of developing an industry on his own," he contended. "I would prefer that our industries be badly administered, but remain ours, than that they be well run and belong to foreigners."[57]

Several Deputies and Senators supported the bill, and their arguments were usually based on the addresses of Juarez Távora and Odilon Braga. Such supporters, usually in the União Democrática Nacional (UDN), tended to represent the more traditional industrial groups, orthodox and internationalist in their economic views, that had opposed Vargas's forced industrialization under the Estado Nôvo. For example, UDN Deputy Pereira da Silva charged that the "O petróleo é nosso" campaign was Communist-inspired; "agents of Stalin" were clouding a question which should be discussed unemotionally by experts, not in the streets. The Communists, he believed, intended to block a realistic solution to Brazil's petroleum problems.[58]

One of the most compelling arguments in support of the bill came from Raimundo D. Padilha (UDN, São Paulo) in a speech in the Chamber of Deputies. To create a state-monopoly company for Brazil's oil, which would also include making Brazil self-sufficient in crude-oil production, was a ridiculous aspiration in his opinion. To substantiate his charge, he estimated how much it would cost Brazil to install a self-sufficient, integrated petroleum industry, using data from the United States (probably the most reliable available). His final figure was Cr$12.5 billion ($625 million), and it is revealing to see how he arrived at it.

In the United States, he discovered, costs in the oil industry were apportioned in this way: production 59.3 per cent, refining 18.3, transportation 12.4, distribution 9.0, other 1.0. *The Petroleum Data Book* (1947) estimated the construction cost of a refinery to be $1,000 per barrel of capacity. A modern refinery of 55,000 b.p.d. would thus cost $55 million, or Cr$1.1 billion. Thus, applying this cost proportionately for a complete industry:

Production	Cr$3,564,480,904	59.3 per cent
Refining	1,100,000,000	18.3
Transport	745,355,196	12.4
Distribution	540,983,610	9.0
Other	60,109,290	1.0
	Cr$6,010,929,000	100.0 per cent
Add:		
Liquid capital	Cr$643,313,600	
(cost per barrel of refining, based on U.S. figure of $1.598, or Cr$31.95, of above refinery).		
Cost of training 300 technicians outside Brazil, at $300 per month	90,000,000	
	Cr$6,744,242,600	

In 1947 Brazil consumed 16.5 million barrels of crude oil. The Recôncavo wells produced an average of 47,450 barrels each for the year. Since the U.S. average was only 4,000 barrels per well annually, however, Padilha suggested that Brazil could reasonably expect 20,000 barrels per well annually in future. With this figure in mind, Brazil would need 825 producing wells to achieve self-sufficency in crude production. Padilha reckoned that 80 per cent of wells drilled are dry (as a rule of thumb, and rather a low one at that), and that a well costs an average of $86,000 to drill. Thus Brazil would have to drill 4,125 wells, 3,300 of which would be dry, at a cost of $354,750,000, or Cr$6,095,000,000.

Thus, the total cost of making Brazil self-sufficient in crude petroleum (to 1947 levels), slightly more than self-sufficient in refining, and creating an integrated industry, would be:

$$\frac{\text{Cr}\$6,744,242,600}{6,095,000,000}$$
$$\text{Cr}\$12,839,242,600$$

(Padilha's figure was Cr$12,420,246,600; he had not reckoned the cost of drilling 825 producing wells, only 3,300 dry ones.)[59] Even though Padilha's figures are correct, based on the construction cost per barrel of a refinery, it is debatable how applicable such a breakdown was to the particular conditions of Brazil. To start at the basis of Padilha's analysis, a refinery would cost more to build in Brazil than in the United States, since virtually all materials would have to be imported. Furthermore, the unique conditions of Brazil (or of any other region in the world, for that matter) would dictate a different distribution of expenditures than in the United States. By the same token, the cost of drilling a well in Brazil — at least anywhere away from the coast — would be higher than the U.S. average, in all likelihood, because of the more difficult terrain encountered, lack of materials and skilled technicians, and so on.

But to examine even the earliest stage of the industry, exploration, raises doubts about the applicability of U.S. figures to Brazil. For example, in 1950 Lewis W. MacNaughton, of the exploration company DeGolyer and MacNaughton, said that a consideration simply of the surface geology of both the United States and Brazil revealed similar petroleum prospects for each country. But a scientific prediction of petroleum resources could not be made without knowing the "third dimension", the subsoil, for which extensive drilling was necessary. It was possible to do this in the United States because there were 8,000 active geologists there, at least half of whom were operating in the field of petroleum. Such was not the case in Brazil, since there did not then exist a single university course in geology (not to be confused with mining engineering, which was taught). "Almost all" those who practised geology in the country were self-taught, and they numbered only "a few dozen".[60]

We have already seen that the cost of handling oil (storage tanks, pipelines, pumps) was higher in Brazil because of its high paraffin content, which necessitated periodic steaming and cleaning of such equipment, which in turn meant either interruptions in production rhythm or excess equipment. Thus, Padilha's figure was almost certainly very conservative.

Padilha closed his speech with the following assessment:

> To summarize, this is the Brazilian situation: low national revenue because of deficient productivity; sparse productivity for lack of financial and technical instruments to develop it; lack of these instruments, because the country has a steel industry with iron, but without fuel of entirely national production, and does not possess petroleum and coal to supply its transportation system.
>
> Thus we are incapable of developing our resources of culture and projects of social service, whether by initiative of the Government or by private means: this is plainly reflected in the indices of illiteracy, in the reduced ability of the elites and painfully in the demographic figures of malnutrition, with particular reference to rural areas.
>
> In these conditions of economic incapacity, the proponents of absolute monopoly intend nothing less than to attribute to the state — that is, to a national company — the financial responsibility of paying for the reconnaissance, the exploration, the development, the transportation, the refining and the commerce of petroleum, with all the technical and administrative problems involved.[61]

The candor of Padilha's speech was rare in the oil debate, both inside and outside Congress. But he believed, with most Brazilians, that his country had enough petroleum not only to become self-sufficient but to meet future increases in demand. This was the "blind spot" Odilon Braga had discussed in 1936.

But then, how could Padilha or any other layman have known otherwise? Little was yet known about Brazil's petroleum geology, because the CNP budget was meagre, and it concentrated on production rather than exploration; as has been mentioned before, the agency's reports did not circulate widely and it did not have a public-relations budget.[62]

Meanwhile, on May 10, 1948, as the debate over the Petroleum Statute bill continued, President Dutra sent to Congress a comprehensive economic development scheme called the SALTE plan. Under the energy sector, the plan set up a four-year program for petroleum on the assumption that intensive exploration would uncover reserves of sufficient size to raise production to 50,000 b.p.d. (estimated consumption for 1948). The plan allotted Cr$1.15 billion ($57.5 million) for exploration; Cr$650 million to build a refinery of 45,000 b.p.d. capacity and to double the capacity of the Mataripe (Bahia) refinery (still under construction) to 5,000 b.p.d.; and Cr$700 million to buy fifteen 15,000-ton tankers to carry all the oil required for national consumption. The plan also permitted the operation of private refineries within Brazil, financed by national capital. (The planners expected that these refineries would handle the extra 30,000 b.p.d. by which consumption was expected to increase by 1952.)[63]

Before examining the political justification for the plan—in so far as oil was concerned—it is worth noting some of the calculations upon which the above figures were based.

Planners estimated, for example, that self-sufficiency in refining capacity would save Brazil $1 per barrel of oil consumed, since the CIF-Rio price of a barrel of "refined products" was $3.80, and of Venezuelan Quiriquiri crude $2.80. Presuming that only cruzeiros—not dollars—were spent on the refining process, and that the same barrel of "refined products" would be produced to substitute for the imported one, such a *foreign-exchange* saving would result.

Further, the CNP estimated the operating cost of a refinery at 70c per barrel, leaving 30c as profit. This cost figure does not jibe with Padilha's (indeed, it is less than half his $1.598 per barrel), which was based upon U.S. figures. We have already examined how low that figure probably was, given the difficulty of handling Brazilian oil. Since Brazil had no refineries worthy of the name operating at the time, and thus no figures upon which to base a realistic cost estimate, the discrepancy between the two estimates is perhaps understandable.

Congress, however, quickly passed the measure and approved the funds requested for it. The reasons for its speedy passage are not difficult to discern. In the first place, it was well timed: the duration and bitterness of the debate over the Petroleum Statute bill had by then raised the distinct possibility that the proposal would be defeated or allowed to die in Congress. Valuable time was being lost—consumption was rising and the tiny Brazilian oil industry was at a standstill—and the SALTE plan appeared to offer a quick solution to the rapidly increasing demand for petroleum. Then, the solution was safe and nationalistic, turning away from liberalism to the well-remembered government control and initiative, while still allowing private capital significant participation in refining. (The plan was thus politically attractive, catering to the wave of nationalism sweeping the country, while throwing a sop to private capitalism.) Further, it promised to effect foreign-exchange savings through refinery and tanker construction, and the massive reserves of hard currencies Brazil had accumulated during the war suddenly appeared to have decreased significantly. Finally, the bill told Brazilians what they believed: that they would find all the oil they needed, soon; all that was required was a little more money and effort.[64]

Now things began to happen. In mid-1949, under the SALTE plan, the government bought equipment from France to build a refinery of 45,000 b.p.d. capacity, using frozen credits. The original site was to have been Belém, on the mouth of the Amazon River, but it was subsequently built near São Paulo.[65] Later in 1949, after years of negotiations and delay, the National Security Council approved applications by two Brazilian groups to construct refineries. One consortium, Refinaria de Petróleo do Distrito Federal, led by Drault Ernani, was permitted to build a refinery of 10,000 b.p.d. capacity at Manguinhos on the outskirts of Rio de Janeiro. It planned to purchase its plant in the United States, and had hired the Foster Wheeler Corporation to erect it. The other consortium, Refinaria e Exploracão de Petróleo União, under Soares Sampaio, was to build a 20,000 b.p.d. refinery at Capuava, near São Paulo. Its equipment would come from Czechoslovakia, with planning and research by Hydrocarbon Research Inc.[66]

Both groups had financial problems and sought aid from the

Bank of Brazil. Some members of Congress objected to public aid for such private concerns, and followers of the "O petróleo é nosso" campaign took up the issue. Some radical nationalists charged that the two groups were front organizations for "the trusts" and asked why a government-owned bank should help foreign interests infiltrate the Brazilian industry. But Dutra, in the spirit of the SALTE plan, permitted the Bank of Brazil to finance the companies and the two projects were subsequently approved.[67]

Early in 1950 a Brazilian government commission signed contracts for a total of nineteen ships in Sweden, Great Britain, and Japan.[68] Unfortunately, the SALTE plan was never fully enacted; it fell victim to bureaucratic wrangling immediately, and became entangled in financial difficulties—budget cuts, delays in receiving appropriations, inflation—after only a year of operation. Its funds and personnel were then trimmed sharply, and it became simply an agency to allocate its own funds to other agencies on a priority basis.[69] But it did stimulate the transportation and refining sectors of the Brazilian petroleum industry, which would alleviate the growing drain on foreign exchange. The plan also provided funds for increased exploration: in 1950 the CNP drilled thirty wells, whereas in 1949 it had completed only seventeen. Production had increased from 300 b.p.d. in 1949 to 930 b.p.d. in 1950.[70]

By the end of President Dutra's term, then, the Petroleum Statute and the liberal approach to development it represented were dead issues in Brazilian politics. A wave of public reaction, in sympathy with economic nationalism, had removed it as an option for policy-makers. Two groups could claim credit for having awakened — and significantly influenced — the public: those army officers who supported Horta Barbosa's thesis on oil development, and the radical nationalists—civilian and military —who staffed the Petroleum Centre. The first group sent copies of Horta Barbosa's speeches under a Military Club letterhead to everyone they judged capable of influencing public opinion (military officers, politicians from all levels of government, media representatives, union members, and teachers at all levels). The second, employing more aggressive tactics, brought the campaign into the streets, adding a mass base to the work of the first group. Within a year of the campaign's inception (i.e., toward the end of 1948), however, the Petroleum Centre had become tarred with Communism, as attacks by Centre orators became exclusively anti-American.[71] But by that time the Petroleum Statute was dead; it remained to be seen whether nationalism had also become identified with Communism, and whether it would lose significant numbers of adherents in the years ahead.

Thus, the Petroleum Statute had been defeated by radical nationalism, operating from the Petroleum Centre, and by emerging developmentalist nationalism, a primitive form being articulated by Horta Barbosa and his corporatist disciples. The campaign struck a responsive chord among Brazilians because the Petroleum Statute was an unpopular bill, and under the postwar political system they could openly register their dislike to some effect. In the first place, the bill was unpopular because it ran counter to the nationalist sentiment of the times. Whereas in the 1930s and earlier, nationalism had been proscriptive and chauvinistic, it became more dynamic during the Second World War. Getúlio Vargas used Brazil's wartime profits to pursue a program of industrialization based upon nationalism. The more liberal economic policy adopted after the war by the Dutra Administration may have resembled a regression to the colonial economy (before 1930) when the nation exported raw materials, imported manufactured articles, and depended absolutely upon foreign markets. There had been too much nationalistic rhetoric under Vargas to revert to this situation.

Moreover, Brazil had a "national investment", however small, in its petroleum. Orators in the late 1940s made much of the fact that oil had been discovered by Brazilians after "the trusts" had said that Brazil had no petroleum. Thus, legislation which appeared to hand over the fruits of this effort to the very foreign groups who had earlier ridiculed it was bound to encounter opposition. The campaign of the Petroleum Centre and of other groups associated with the "O petróleo é nosso" theme struck a responsive chord with the public and with Congress.[72]

Finally, the Petroleum Statute bill fell prey to political faction. Upon being forced into political exile at his São Borja ranch in 1945, Getúlio Vargas formed the Brazilian Labour Party (PTB) to organize his personal following. By December 1946 he had broken openly with the Dutra government, and had begun to prepare for the 1950 presidential election. Vargas's wide support and his deft political manoeuvring apparently assured his election; his *"trabalhismo"* creed and his record while President from 1930 to 1945 made it certain that he would not support liberal measures such as the Petroleum Statute.[73] Indeed, in a campaign speech in 1950, Vargas viewed the bill as a manifestation of the "new wave of pessimism and discouragement" which had penetrated to "the highest official circles". The Executive itself had believed, Vargas charged, that Brazil could not develop its petroleum resources on its own. Emphasizing that the country which did not control its petroleum could not be considered independent, the ex-dictator recalled that Brazilian petroleum had been a reality since 1939. While he endorsed the principle of hemispheric solidarity, he saw no reason why Brazil should sacrifice the benefits from the exploitation of its petroleum. He denied that he wanted to exclude foreign capital; what he did not want was the alienation of natural resources to foreign companies. He concluded:

> Let us speak clearly: that which is indispensable to national defence, which constitutes the basis of our sovereignty, may not be handed over to foreign interests; it must be exploited by Brazilians through organizations predominantly Brazilian, and, if possible, with a high percentage of state participation, avoiding in this way surreptitious penetration by threatening monopolies.[74]

Vargas's statement must have led his listeners to presume that Brazil's flirtation with a reasonably liberal oil policy would end with his resumption of office; that the country was to return to somewhat more nationalistic precepts.

4
ECONOMIC NATIONALISM TRIUMPHANT, 1950–1954

When Vargas returned to the Presidential Palace in 1950, it was not at the head of a popular wave of nationalistic fervour, but as the result of a number of personal deals with various powerful political bosses from every major party, even the anti-Vargas UDN. He had thus emerged from the election a pragmatic politician, committed merely to increased state intervention to foster national development. The politicians to whom Vargas owed favours were committed not to ideologies but to winning power to dispense patronage. They would support Vargas as long as he could keep public opinion behind himself. They spoke for both old and new political forces: traditional rural groups and populistic urban elements. And it should be remembered that Vargas's strongest role as dictator had been that of mediator or conciliator between antagonistic political forces, time and again snatching compromise from seemingly irreconcilable positions.[1] So it is not surprising that Vargas proposed a compromise solution for the thorny and divisive problem of the development of national oil resources: to honour his debts to groups with different positions on oil, to break the development bottleneck which controversy had created, and in the process to re-establish his dominance over the Brazilian political system by reconciling apparently antipathetic groups.

Vargas, however, misread the political climate of Brazil in 1950 and overestimated his own ability to fashion an acceptable solution to the energy crisis. Brazil had changed since 1945, as the unique "O petróleo é nosso" campaign had shown, and Vargas clearly had not grasped that fact. In the first place, Vargas failed to notice a growing aversion within the military to political activity. While the majority of officers had privately endorsed Horta Barbosa's stand on the Petroleum Statute, and many of them had been sufficiently stirred to make public statements of support for it, the association of a few of their number with the Communist-tainted Petroleum Centre had been embarrassing. By 1950 a growing minority of officers — many of whom had actively supported Horta Barbosa — had begun to feel that the military should restore its integrity by public silence on contentious matters.

Upon his election late in 1950, Vargas announced plans for a broad development program which would galvanize both the public and private sectors (in Skidmore's terms, developmentalist nationalism). He looked to the more nationalistic wing of the military for support, just as he had used the economic nationalism of the same group in the industrialization of the Estado Nôvo. Unfortunately, the men to whom he again instinctively turned for support were under attack from their more moderate colleagues and fast losing power by the time Vargas reached office.

In May 1950, however, the Clube Militar elected a nationalistic slate, headed by Generals Estillac Leal and Horta Barbosa, to the club's directorate. Immediately, the influential and prestigious organ of the club, the *Revista do Clube Militar*, adopted a tone of militant nationalism and became, in the process, stridently anti-American. The peak came in two months when a July article charged that the United States was responsible for starting the Korean war.[2] The outcry within the military was substantial, especially from the group of officers who had served in the Brazilian Expeditionary Force in Italy during the Second World War (and had grown to admire the U.S. military from close association with it) and who were now associated with the increasingly influential Higher War College. A campaign within the club against Estillac Leal and his politically minded supporters began immediately, and it was soon clear not only that Estillac would be turned out of office in 1952 (accompanied by Horta Barbosa), but that the military would stop participating in public debates. Thus, Vargas found, by the time he took office, that not only was his support within the military ebbing sharply, but he was powerless to stop the process. He appointed Estillac Leal War Minister, but this further alienated the bulk of the officer class. Vargas had suddenly shown himself to be politically ineffective.³

Vargas also misjudged the civilian political climate. He underestimated the effect of the "O petróleo é nosso" campaign, not realizing that it had removed the possibility of foreign participation in the development of national oil resources. Two conclusions are possible: either Vargas simply did not realize to what extent the Brazilian public had become exercised over the implications of the Petroleum Statute, and thus erred as a politician by introducing a bill that would permit a modicum of foreign investment; or he believed that the Brazilian political system could be manipulated as it had been during his dictatorship, and that he would have a free hand to effect a compromise between contending interest groups, with merely a passing nod to public opinion.

At any rate Vargas believed, in company with supporters of the Petroleum Statute, that Brazil's most urgent need was to develop the oil that virtually no one doubted was in the national subsoil. The solution he proposed was in large part an elaboration on the defeated statute, with substantial revenues from a variety of sources, including foreign private capital. He advocated the creation of a vertically integrated state-controlled company, patterned on the organization of the large international oil companies; but instead of revenues from concessions as the statute had proposed, he recommended that they come primarily from taxes on oil products.⁴

Brazil's urgent need to develop this crucial energy source could not have been overstated in 1950. Indeed, by 1949, Brazilian officials were aware that the country faced a fuel crisis, exacerbated by dwindling foreign-exchange reserves. Over one hundred thousand motor vehicles were imported in 1951, and the figure would increase. Basic industries could not expand without substantial increases in energy, and oil was expected to supply most of that need. Vargas planned to modernize the railroads by introducing diesel locomotives. Fuel imports had increased 20 per cent per year since 1949: in 1951 to import consumption requirements of 120,000 b.p.d. cost $200 million; by 1955, requirements were expected to be 250,000 b.p.d., costing $500 million (which had to be paid in scarce dollars).⁵

The Vargas government faced the prospect of greater and

greater foreign-exchange allocations for petroleum imports, foreign exchange which would for the immediate future be earned almost completely by coffee sales (the price of which was subject to fluctuations on the world market). Future balance-of-payments crises might well cause fuel shortages which would slow economic growth, and perhaps even nullify attempts at modernization. And Vargas was determined to modernize the Brazilian economy. He wished to build on his success—during the Estado Nôvo — in creating a national steel industry, and fashion a self-sufficient economy for the nation, removing it from absolute dependence upon a capricious foreign market. Reliable national sources of energy were vital to such a plan, and so Vargas decided to allocate most of his government's limited resources to the development of oil and electricity.[6] In such a situation, and faced with such prospects and goals, Vargas could be excused for either ignoring or underestimating the strength of the nationalistic campaign to exclude foreigners from participation in oil development. Faced with such massive revenue requirements, Vargas would naturally want as many income sources as possible, and would wish to keep open the possibility of some help from the outside.

Immediately after his election, Vargas appointed an advisory board, the Assessoria Econômica da Presidência da República, staffed by brilliant young economists and technocrats. He then directed it to draft, as part of a comprehensive plan for modernizing the national economy, an oil law which would create a company to stimulate oil-industry growth, with revenues chiefly from taxes on oil products but from private sources — national and foreign—as well. Early in May 1951, before the new bill appeared, executives of Standard Oil of New Jersey visited Brazil, and said that their company was "ready to participate on an equal share with Brazilian capital in the exploitation of Brazilian oil resources". Then, Foreign Minister João Neves de Fontoura told a joint congressional committee that the United States was ready to help Brazil exploit her oil reserves. He charged that the Petroleum Centre, "a Communist organization, the mouthpiece of the Kremlin in this country, is bent on thwarting all our efforts in producing petroleum."[7] It appeared that the new government was preparing to permit foreign participation in the national petroleum industry, perhaps to the degree which would have been allowed under the Petroleum Statute (a possibility which Vargas had entertained during the Second World War).[8]

As Vargas and his advisory board worked on a new oil proposal, the CNP recorded some modest gains. In 1951 its Mataripe refinery began operations, and work immediately began to double its capacity to 5,000 b.p.d. The refinery at Cubatão, São Paulo, purchased under the SALTE plan, began to take shape along with an adjoining fertilizer factory. Crude-oil production reached an average of 1,620 b.p.d., which the CNP estimated could be 20,000 b.p.d. In its 1951 report the CNP lamented the "deficiency of technical and financial resources".[9] In the same year, steps were taken to alleviate one part of that complaint: the University of Brazil agreed to offer courses to train technicians for the oil industry. The CNP lost João Carlos Barreto, its president since 1943, and under his interim successor, Plínio Cantanhede, the agency became somewhat more nationalistic: a proposal to build a 30,000 b.p.d. refinery across the bay from Rio de Janeiro was rejected; nationalists claimed financing was from Socony–Vacuum Oil Company, Inc., and the Brazilian proponents were mere figureheads, "Trojan horses" to permit the entry of "the trusts".[10]

On December 6, 1951, Vargas sent the new proposal for a comprehensive oil law to the Chamber of Deputies, accompanied by a long descriptive message. The legislation created a mixed company, Petróleo Brasileiro Sociedade Anônima (Petrobrás), which was to have a monopoly over all phases of the national oil industry except distribution (presumably because scarce financial resources were to be allocated to new industries, so distribution was left in private, mainly foreign, hands).

In his speech accompanying the bill, Vargas called upon "the financial and human resources" of Brazil to reduce dependence on oil imports. He stressed that both the national consumption of such products and their proportion of total imports were increasing and predicted that oil consumption would double by 1955, imposing a crippling burden on foreign-exchange reserves. Vargas cited figures to lend weight to his predictions: in 1939 oil imports had been 7 per cent of imports, by value; in 1946

7.6 percent; by 1950 11.3 per cent; and 1951 would see them surpass 13 per cent of total imports. Consumption had increased 6.4 per cent annually in the period 1931–40, and 11.9 per cent annually from 1941 to 1950; the period 1946–50, however, had seen a 19.5 per cent mean annual increase, and between 1949 and 1950 consumption of petroleum derivatives had jumped a staggering 22.3 per cent. The fact that consumption would double from 1950 to 1955 meant that the refineries planned by the Dutra government would be serving only half national consumption within five years. Petroleum, he emphasized, was the "basic factor for the economic emancipation and social wellbeing of our people". If the growing national demand could not be met, Brazil's economic growth would be curtailed. Speed was of the essence.

The bill was faithful to the nationalistic spirit of existing oil legislation, Vargas said, and the proposed company, Petrobrás, would be genuinely Brazilian, with national capital and administration. Any danger of control passing to national or foreign private capitalists had been forestalled by limiting the acquisition of voting stock, by the fact that the President of the Republic chose the president and executive directors of Petrobrás, and by the requirement of a presidential decree to change any clause in the law.

Vargas listed six sources of revenue for the proposed company. The federal government would subscribe the initial capital, Cr$4 billion ($200 million), and would thereafter retain a minimum of 51 per cent of the shares. (The project provided that total capital would increase to Cr$10 billion by 1956.) Secondly, Petrobrás would receive part of the tax on liquid fuels and part of the duty on automobiles (all of which were at that time imported); both these imposts would be raised. A third source would be a new tax on luxury articles. Part of the revenue from state and municipal taxes on liquid fuels would also go to Petrobrás, and owners of automobiles would contribute through compulsory purchase of "certificates", which could subsequently be exchanged for shares in the company. Finally, Petrobrás would gain revenue from the sale of stock to the public; shares could be purchased by Brazilian citizens, by companies organized within the country (with no stipulation as to the nationality of their employees), and by public agencies and the various levels of government. Vargas emphasized that the bulk of the private capital would be subscribed by vehicle owners: those who had a direct interest in the company's operations.[11]

The Petrobrás project differed from the Petroleum Statute in that it gave the government much more control over the petroleum industry and immediately provided substantial capital to the new industry from sources outside its operations. The Petroleum Statute had provided that the government, through the CNP, could delegate all the functions of the industry (except distribution) by means of authorizations or concessions. By contrast, the new proposal created a company, controlled by the government, to perform the same functions directly; there was no mention of concessions. Moreover, under the Petroleum Statute, financing could be entirely from holders of concessions or authorizations, whereas under the new measure the government provided the major part of Petrobrás's capital.[12]

Looked at dispassionately, Projects 1516/51 and 1517/51 (Vargas's Petrobrás bill and its companion setting out the company's revenue sources) should have appealed to a broad spectrum of the population. In the first place, not only was there a severe restriction on foreign investment in the industry, but control would remain in the hands of the federal government, with further shares allotted to state and municipal governments. This degree of central-government control should have pleased the many Brazilian corporatists, and have been sufficient to allow radical nationalists not to worry about the modicum of foreign participation.

Then, too, financing for the company appeared to be quite adequate for its task. The amount of money initially allotted to the industry, as well as the amount it was assured of in the future, seemed adequate to find, extract, refine, and transport enough oil within the national territory to serve projected needs until the industry could finance itself—that is, providing there *was* enough oil in Brazil even to fill projected needs, let alone to allow the company to pay its own way. But then it is unlikely that many Brazilians—or any at all—doubted that their country was rich in oil.

Financing also seemed to have been well planned. Existing and potential sources of revenue had been harnessed with imagination and foresight to give Petrobrás a sound base from which to operate. Indeed, the care with which financing had been laid out, and the part—however small—private industry was exhorted to play in it, indicate a hope on Vargas's part that he could win supporters of the Petroleum Statute over to his bill.

Finally, Vargas adopted tones of nationalism, unity, and urgency in his speech presenting the bill; he was clearly trying to harness the campaign which had defeated the Petroleum Statute. Perhaps, by stressing the shortness of time, he hoped the bill would pass without a prolonged fight over the role allowed foreign investors. Or perhaps, as seems more likely, he wanted to get the company going as quickly as possible because it was the basis of his development program, and simply did not expect anything like the negative reaction his bill aroused. In other words, he misjudged the political climate.

So the congressional debate resumed where it had subsided in 1950. As it progressed, the Chamber of Deputies tended to represent the "masses" and the new industrial middle class, and the Senate the more traditional business and aristocratic classes. The former house, which was to have the last word before the bill was returned to Vargas, was inclined to favour rigid state control, while the latter was anxious to preserve a role for private enterprise.

Euzébio Rocha (PTB) initiated discussion for the radical nationalists and proposed, on January 28, 1952, a substitute plan which would exclude private capital from all phrases of the industry except distribution.[13] Lôbo Carneiro (PTB) pledged the support of the Petroleum Centre for this bill and read a statement from it charging that Vargas's bill was not nationalistic but "constituted, in truth, the long-awaited opportunity for the foreign trusts—especially Standard Oil—to penetrate the domain of the exploration and industrialization of national petroleum."[14]

The statement, signed by the Centre's Acting President General Felicíssimo Cardoso, correctly pointed out that subsidiaries of "the trusts", if organized in Brazil, could purchase shares in Petrobrás. Moreover, sufficient ambiguities existed in the bill to permit "the trusts" to seize control of Petrobrás; for example, the certificates given car owners could be converted into shares, but there was no stipulation on the nationality of the owners. The statement concluded that such ambiguities represented an attempt to confuse public opinion to facilitate infiltration by "the trusts". The Centre called upon "all true patriots" to repel the attempt and "to intensify the campaign in defence of the threatened fatherland". Then the Petroleum Centre issued a further statement, reiterating its support for Euzébio Rocha's substitute bill, but calling for amendments to it which would put the distribution of derivatives under the state monopoly, annul the contracts for private refineries, and nationalize the sale of imported petroleum products.

To the charge that the nation had insufficient capital to create a company such as Petrobrás, with all the operations it would have to perform, Lôbo Carneiro replied that it was "only Standard Oil propaganda", which honest people did not believe.[15] Manhães Barreto (PSP),[16] speaking in support of the government bill, dissented from his radical colleagues, warning them that the oil problem was too serious to be discussed in the "demagogic realm of romantic suspicion", subject to the "peevish eruptions of our verbal adolescence". We may not have money, he said, "but we do have phrases." Radical nationalists had likened the emancipation which would result from the native exploitation of Brazilian oil to the abolition of slavery. Barreto observed that when the slaves had been liberated no one had any idea of what to do with them. "We satisfied the poets and demagogues," he asserted, but the move had run counter to contemporary economic reality. The petroleum problem could not be resolved in the manner of other historical movements in Brazil, which he pictured as "flags unfurled amid a lack of awareness of inherent reality". Petroleum should be discussed objectively, not poetically. Brazilians could not say the oil was theirs until they were sure they had the resources to develop it. A restrictive policy without this assurance would mean the ruin of the national economy.[17]

The debates produced surprising shifts in party alignments, when early in May 1952 the UDN, which had traditionally held a relatively open attitude toward foreign investment and was the party of such men as Odilon Braga and Juarez Távora, swung to state monopoly. The UDN National Directorate made the decision at an extraordinary meeting (Braga and Távora, among others, voting against it). Then the party declared its support for the government's policy on petroleum, but called for a state monopoly (with no private-capital participation) over the exploration, exploitation, refining, and transporting of oil, and a state company to perform those operations.[18] Clearly, full state control had captured the popular imagination, and the UDN had responded to the mood of the electorate.[19]

Bilac Pinto, a leading spokesman for the UDN in the Chamber of Deputies, said the party had decided that state monopoly was the best policy for Brazilian oil development at that time. He claimed that changes of policy were normal in public life; Getúlio Vargas had done the same, advocating state monopoly while in exile in Rio Grande do Sul and then calling for a mixed company "once elected". This, Bilac Pinto contended, was a far more serious change than that just effected by the UDN.[20]

The last point makes sense when one considers that the UDN had come into existence as an anti-Vargas coalition. It sheltered such implacable foes of Vargas as Carlos Lacerda, whose campaign in the media against Vargas was becoming increasingly vitriolic and would ultimately help to bring about Vargas's suicide. In such a climate, the fact that the UDN also sheltered a number of such proponents of a liberal oil policy as Távora and Braga meant little. The important task was to bring Vargas and his government down. An oil bill in keeping with its apparent economic philosophy, along the lines of the Petroleum Statute, clearly would have been defeated in Congress, thus helping Vargas. So the UDN joined the radical nationalists, trying to defeat Vargas's bill on the grounds that it was not sufficiently restrictive!

It is possible that within the UDN the supporters of a liberal oil policy believed that something like the Petroleum Statute could be pushed through after Vargas's bill had been defeated. Certainly Vargas wanted as liberal a bill as he could get, in order to tap the widest possible range of revenue sources; so those members of the UDN might have believed Vargas would support their proposal rather than a more restrictive one. But such an exercise must have been based on the belief that the "O petróleo é nosso" campaign had died, and that its adherents had somehow come to their senses. Subsequent events would demonstrate that the campaign was by no means dead.

It must be stressed, however, that the espousal by the UDN of Petrobrás—albeit as a complete state monopoly—signalled the end of free enterprise as a possible means of exploiting Brazilian oil reserves. The UDN clearly recognized that a liberal oil policy had become untenable in the contemporary political climate; the proof lying in the party's decision to oppose Vargas, their enemy, from a radical-nationalist stance rather than from a liberal one. Indeed, development of Brazil's oil by private companies ceased from that point to be a political possibility.

The other major parties were less specific. The PTB, since it owed its existence to Vargas, tended to support his bill, but its members included such radical nationalists as Euzébio Rocha and Lôbo Carneiro, who campaigned against the administration. The PSP similarly tended to support Vargas on the issue but made no policy statement. The Social Democratic Party (PSD) had no declared policy; a few of its members swung with the UDN to state monopoly, but most supported Vargas.[21] Vargas himself appears to have kept aloof from the debate in Congress, save to urge the politicians to speed their deliberations. In one of his very few speeches on the subject (in Candeias, Bahia, near which is located a major oilfield), he emphasized that a mixed company would be the best for development of national oil reserves.[22]

Of the nationalistic groups campaigning outside Congress, the Clube Militar and its review were suddenly silenced in May 1952. The dispute between radical nationalists and moderates within the club had grown so bitter, and the former group had lost so much influence, that in March 1952, moderates forced General Estillac Leal to resign as both War Minister and president of the club. Estillac's resignation also came because his relations with a leading anti-Communist general, Zenóbio da Costa, had deteriorated, and because he had been unable to maintain "discipline" among the officer corps. This latter failure meant that the officers had lost confidence in Estillac's ability to resolve such problems as salary differentials and deficiencies in equipment. The subsequent elections for the club directorate in

May saw the moderate candidates defeat the radicals by a margin of two to one (a critical loss of support for Vargas). A few days later General Alcides Etchegoyen, the new president, said the subject of oil was no longer of "internal interest" to the organization or its review. "I was elected by the military, who do not want such a subject discussed any more in the club," he told a questioner. The new president represented the wing of the military which supported Juarez Távora on foreign investment.[23]

In June 1952 the Petroleum Centre made plans for its Third National Convention of Petroleum Defence in July, and organized a nation-wide series of lectures and debates in preparation for the event. The Brazilian government revealed in June that the United States Secretary of State was coming to Brazil in July, and the political police asked the Petroleum Centre to postpone its convention until August. Antônio Corrêa said in the Chamber of Deputies that this request made his government appear subservient to that of another nation, and the Centre refused to alter the date. Lôbo Carneiro in turn told the Chamber of Deputies that the first such convention had brought down the Petroleum Statute and that the third would defeat Petrobrás. State monopoly had now become the thesis of the Brazilian people, Lôbo Carneiro claimed, and the Petrobrás project would fall before public opinion. Messages of support for state monopoly had been coming from municipal councils to him personally and to the Petroleum Centre for two months, he said, and student unions had been campaigning for the cause. The government had been promoting Petrobrás on radio and in the press, but with no effect upon public opinion; the people knew that "the trusts" could insinuate themselves into the exploitation of Brazil's petroleum reserves through Petrobrás.

Shortly thereafter, Lôbo Carneiro charged that police in various parts of Brazil were persecuting adherents to the state-monopoly thesis. Colonel Benavides, vice-president of the Petroleum Centre, had been harassed while on a lecture tour in the northeast. Police in Teresina, Ceará, had threatened the staff of the newspaper *Emancipação* (linked to the Petroleum Centre), and had seized all the copies of one issue. Lôbo Carneiro also cited cases of arbitrary arrests and beatings, all of which made excellent publicity for the organization and its cause. (Some of the cases of arrest and beating he cited had no connection with the Petroleum Centre or the cause of state monopoly, and Lôbo Carneiro admitted this as he was reciting them; apparently he was adding ammunition to his story.)[24]

Through most of June 1952 the Chamber of Deputies considered 126 amendments to Project 1516/51 (the official designation of Vargas's bill). The first two indicated the range of opinion. Amendment number one put distribution and all refining into Petrobrás's hands; in short, it made Petrobrás a genuine state-monopoly company. In keeping with the UDN's new stance, one of its members moved the amendment. Among its endorsers was Euzébio Rocha of the PTB, lending credence to the theory that Brazil's political extremes were drawing together to defeat the government. The second amendment, moved by a PTB Deputy, gave the CNP a monopoly over the oil industry with authority to make contracts with private firms, national or foreign, to perform the various functions of the industry. The only significant difference between this amendment and the defunct Petroleum Statute was that the former talked of "contracts" where the latter used the word "concessions". Presumably this distinction would give the CNP more authority, since a concession implies loss of control over sections of national territory and over the resource extracted for stated periods of time, while a contract need not imply such a loss and usually includes specific provisions for payment in money or kind for the task performed.[25]

It is difficult to explain why a member of what was reputedly Vargas's personal party, the PTB, should have proposed a liberal alternative to Vargas's own bill. It is conceivable—barely—that Vargas himself wanted the amendment in order to create an alternative acceptable to him should his original proposal fail. (Vargas's Vice-President claimed later that Vargas viewed state monopoly as a "Communist" solution which would keep Brazil's oil in the ground.)[26] Another explanation, perhaps more likely, is that the PTB, like every other Brazilian party at that time, included within its ranks a host of personalistic politicians who saw parties merely as vehicles to help them take power. Ideologies were incidental baggage to such men.

Of the remaining 124 amendments, the fourteenth was significant: it eliminated the clause in the Petrobrás bill which permitted a token measure of foreign investment. Under the amendment, shares could be purchased only by individuals who were native-born Brazilian citizens or who had been "naturalized for more than five years and residents of Brazil". Further, if qualified citizens were married to foreigners, the marriage had to be of "separate estates", so that shares could not be transferred to the foreign spouse.[27] This amendment ultimately was adopted, and foreign investors were barred from direct participation in Petrobrás.

The amendment appears to have originated with the PTB, which had apparently grown disgruntled over Vargas's neglect of it while he wooed other factions. No one likes to be taken for granted. No less a radical nationalist than Euzébio Rocha supported the amendment, but so also did Osvaldo Fonseca, who had proposed the liberal second amendment (and on the same day!). Clearly, Vargas had some political fence-mending to do.

During June, Lôbo Carneiro in the Chamber of Deputies, Artur Bernardes in the Senate, and other radical nationalists periodically introduced into the record long lists of groups that had sent messages of support for state monopoly. State assemblies and municipal councils in impressive numbers endorsed state monopoly, as did the National Student Union and some state student groups. The number of such messages demonstrated the depth of the state-monopoly movement throughout Brazil.[28]

Vargas seems also to have played an unforeseen role in the generation of this wave of public support by having talked early in 1952 with Euzébio Rocha, the mover of the more nationalistic substitute oil bill also before Congress. His colleagues regarded Rocha as something of an oil expert, and he also commanded a substantial following in Rio and São Paulo, with which Vargas had to reckon. In their conversation, Rocha claimed that Vargas had expressed a strong preference for a nationalistic solution to the oil problem. Rocha then took it upon himself to change the Petrobrás bill to conform more closely to Vargas's true feelings, or what he believed them to be.

It appears that Vargas made a tactical error in seeing Rocha, and that to win his support he represented himself as a stronger nationalist than his own bill implied. To Rocha and his following, Vargas must have seemed poorly advised in the drafting of the Petrobrás bill; some of them even suggested that "the trusts" had penetrated the President's circle of technocratic advisers.[29] The massive outpouring of support for Rocha's and the radical nationalists' state-monopoly solution for the Brazilian oil problem partly arose from this interpretation of Vargas and the Petrobrás bill. If the President had been given bad advice, the rationale may have run, his citizenry were duty-bound to seek to counter that advice in order to save the nation. Hence the petitions, resolutions, telegrams, and letters began to pour in, in all likelihood at the suggestion of the radical nationalists.

Outside Congress, the Convention of Petroleum Defence was held as planned on July 5–8, 1952. Lôbo Carneiro reported that about 600 delegates attended from seventeen states and the Federal District, representing student unions, the army and the navy, state and municipal governments, and various professions. The delegates declared their opposition to foreign-capital participation in the Brazilian oil industry, because "such capital is purely and simply from the great trusts". They also opposed the participation of national private capital, not only to eliminate figureheads, but because the industry was by nature monopolistic and the state should therefore control all its aspects, applying the profits to the industry's growth and to "the collective benefit". The convention repudiated Project 1516/51 as "unpatriotic and of *entreguista* (sellout) nature", and called on the Chamber of Deputies to adopt the first amendment. The delegates called "immediately" for regional conferences to keep the issues before the public.[30]

Throughout the Brazilian winter (July-September) petroleum policy remained a lively issue. Supporters of the government proposal and of state monopoly—by that time the only alternatives feasible in the climate of opinion—used every means at their disposal to sway the public. Proponents of state monopoly charged that their opposition was in the pay of "the trusts", while supporters of the Petrobrás bill called their opponents Communists.[31]

In September 1952 the Chamber voted. Of the various amendments passed, the most important was the fourteenth, which excluded foreign capital from participation in Petrobrás, even through companies organized in Brazil. The deputies approved the altered project on September 18, and a week later forwarded it to the Senate.[32] Lôbo Carneiro, representing radical nationalism, said he was "bitterly disappointed" in the Chamber of Deputies. He termed the vote "an affront to the Brazilian people", who, in more than four years of civic campaigning, had called "in an unequivocal manner"for a state monopoly.[33] (The bill which the Chamber of Deputies passed left distribution in private hands and reserved a share of the refining sector for private companies.) On the other hand, Raimundo Padilha (UDN), a strong proponent of foreign participation in national oil development, called the passage of the Petrobrás bill "a Communist victory".[34]

Majority public opinion by mid-1952 was probably best represented by an article by Professor Maurício Joppert in the *Revista do Clube de Engenharia*. The author, a UDN deputy from the Federal District (then the city of Rio de Janeiro) and an engineering professor, declared that he was a firm believer in inter-American economic co-operation. "New countries, such as Brazil," he said, "have developed with foreign assistance and still need collaboration from the outside to pursue or induce progress and for the exploitation of their natural resources." But, he said, "petroleum is different": its history had been unusually violent, governments had fallen over it, there had been "secret wars" among nations for it, as well as intrigue and corruption on a grand scale. Because of this sordid tradition—he cited the Mexican experience as a good example—Joppert said he advocated a state monopoly for the exploitation of Brazilian petroleum. "It is the only way to maintain our sovereignty as an independent nation."[35]

Not every Brazilian held this attitude. Much of the business community, for example, wanted a liberal policy toward foreign capital, and most of the São Paulo newspapers echoed this sentiment. *Folha da Manhã*, for instance, epitomized the attitude of Brazilian business when it said, "No country without coal or petroleum is really an economic power in the machine age." What mattered was getting the oil out of the ground and putting it to use, so that the economy could grow. Ownership could be settled later.[36]

A striking feature of the debate over Brazil's oil policy must be reiterated: no public figure questioned the belief that Brazil held vast quantities of oil and that "the trusts" were eager to exploit national reserves. Both *entreguistas* and nationalists believed that a liberal policy would be sufficient to bring in the international companies; they differed only in their view of such a result as a good or bad one.

Meanwhile, a crisis was developing in the Brazilian economy. As more and more foreign exchange was required to pay for the rising volume of oil imports, as well as of other commodities, Brazil was amassing huge debts with supplier nations. (The *New York Times*, in August 1952, noted that Brazil was having to pay close to $200 million *in dollars* that year for imported oil.) At the same time, Brazilian exports were overpriced and, with the exception of coffee (which earned 70 per cent of Brazil's hard currency and for which demand far exceeded supply), were having to be stockpiled in Brazil for lack of buyers.[37] At the end of 1952 the nation had an adverse balance of payments of $162 million. Inflation was occurring at a critical rate: 21 per cent in 1952.[38] But the crisis had little or no effect upon the deliberations of Congress. The nationalist deputies ignored the cost of creating a state oil industry; when they mentioned finances at all, they stressed the apparently vast savings which would result from the industry's creation.

At the same time, the CNP was able to point to some advances. For example, in 1952 production rose to an average of 2,000 b.p.d. The CNP drilled sixty-four wells in that period: forty-four yielded oil and two gas. (There had been little exploratory or wildcat drilling; most activity had been centred on the Bahia Recôncavo, developing known fields.) The Mataripe refinery having begun operations, preliminary studies had already been completed to expand its capacity to 15,000 b.p.d.; the construction of a unit to produce lubricating oils had begun. Construction of the new refinery at Cubatão (authorized under the now-defunct SALTE scheme) had been delayed, but equipment had begun to arrive and the complex would be finished by the end of 1954, in the CNP's estimation. Finally, the agency was proud to note that it was now able to purchase some specialized equipment in Brazil (local industry was responding to the needs of the oil industry); and the Polytechnical School of Bahia had initiated a refinery technicians' course.

In plenary session the CNP also authorized I.B. Sabbá & Company Ltd. to construct and operate a refinery of 2,500 b.p.d. capacity at Manaus, 1,000 miles up the Amazon River. (The next year the CNP altered the authorization in favour of a company Sabbá had incorporated for that purpose, and doubled the capacity of the projected refinery.) It would draw its crude from Peru's Ganso Azul field—just up the Amazon—and serve the Amazon Basin.[39] This brought to four the number of privately owned refineries in Brazil, either in operation or in varying stages of preparation. Projected production of all four was about 45,000 b.p.d., less than one-third of consumption in 1952.

Before the Petrobrás project officially arrived in the Senate, there were occasional speeches in support of various oil-development policies by members of the upper house. For example, publishing magnate Assis Chateaubriand (PSD) advocated development of the industry wholly by private enterprise, national or foreign. He poured contempt upon his nationalist opponents, calling them "Indians"; if their policy prevailed, he forecast, the senators would soon be dressed in loincloths, carrying bows and arrows into the Chamber.[40] It indicates something of the relative polarization of the Chamber of Deputies and the Senate that no member of the former body subscribed publicly to Chateaubriand's thesis, while at least a handful of senators did.

On the other hand, Senator Landulpho Alves (PTB), ex-Governor of Bahia, promoted Vargas's bill, warning of the danger to Brazilian sovereignty from private-capital control of the national petroleum industry. Alves saw oil as "a complex of political, economic and financial problems", upon which depended national "social and economic well-being and tranquility in social relations, based on a full and unrestricted defence of national sovereignty".[41] Plínio Pompeu (UDN, Ceará) espoused what he saw as a compromise: he wanted the jurisdiction of a state oil company restricted to the existing producing zone (the Recôncavo), and the remainder of the country opened to exploration by concession.[42]

Finally, on October 30, 1952, the Petrobrás project was officially introduced into the Senate, where Plínio Pompeu summarized the case for the *entreguistas*: "Our petroleum will remain buried eternally if we do not have foreign assistance." For the nationalists, Kerginaldo Cavalcanti retorted: "It is better that it remain buried for the glory of future generations, than it be handed over to the exploitation of the international trusts." [43]

Assis Chateaubriand, however, was the first member of Congress since before the Second World War to question publicly the basic assumption that Brazil held vast oil reserves. Citing a critical article in *O Estado de São Paulo*, he asserted that "competent authorities" were not at all certain of the petroleum possibilities of the country. Unfortunately, the senator went on to charge that the nationalist campaign had been "inspired and directed by the international Communists". Chateaubriand asserted that the aim of the Communists was to ensure that Brazil would have no oil; then, in the event of an international conflagration, the nation's economy would be paralysed, and a revolt and Communist takeover would result.[44] By thus peremptorily dismissing a movement in which so many genuine nationalists were involved, and which had won such widespread popular endorsement, he weakened his whole statement about national oil resources. Brazilians suspected that foreign interests were sponsoring Chateaubriand's campaign to throw open national oil reserves; his remarks thus did not get the attention they deserved.

By March 1953 the Senate was considering amendments. The nineteenth was significant, calling for the exploitation of national oil reserves by companies organized within Brazil, as well as by the CNP and Petrobrás.[45] Othon Mader (UDN, Paraná), the senator who proposed the amendment, had earlier declared himself one "of those who combat state monopoly".[46] Mader received numerous messages of support for his liberal position and read several of them in the Senate; it is significant that they were virtually all from commercial associations.[47] Clearly, traditional, export-oriented businessmen were against the nationalism represented by Petrobrás, since it would further entrench the newer, industry-oriented business elite.

But the traditional business community lacked an effective voice in the Congress. Although the Senate tended to represent such business and financial groups, its amendments were subject to approval or rejection by the Chamber of Deputies, which responded to the newer business philosophies.[48] Not even on the local level did political power lie with traditional business and commerce, since state and municipal legislatures apparently opted in favour of state monopoly. As Plínio Pompeu noted, "the masses are at the side of those who defend 'O petróleo é nosso': while the elites permit the collaboration of foreign capital." He suggested that the duty of political leaders (meaning the senators) was to "infiltrate the masses" and "steer them from the wrong path".[49]

Landulpho Alves responded to Mader by reading messages of support for the state-monopoly position. He and Kerginaldo Cavalcanti (PSD, Pernambuco) called the groups who had sent messages to Mader the "conservative classes", who desired a high cost of living and free initiative in industry and commerce and whose principal motivation was to sell staples to the poor at prohibitive prices. Alves further charged that Mader's supporters were "incapable of forming opinions on great issues".[50]

Of the various committees which reported to the Senate, it is significant that the National Security Committee saw no threat in the original Petrobrás project. As far as the General Staff was concerned, said the committee, even the Petroleum Statute had presented no security problem. What had interested the General Staff then, and still concerned them, was the rapid development of a ready national supply of petroleum. The report was made February 10, 1953. The *relator* of the report was Senator Ismar de Góes (PSD), who advocated a solution similar to the Petroleum Statute; his report reflected this thinking.[51] The National Security Committee in the Chamber of Deputies, on the other hand, had a nationalist presiding over it, and it recommended adoption of Euzébio Rocha's restrictive substitute.[52]

The Senate voted early in June 1953 and returned the bill to the Chamber of Deputies with thirty-two amendments, several of which would have made the Petrobrás project slightly more liberal than in its original (1951) form. The Senate amendments, for example, allowed foreigners to buy unregistered shares in Petrobrás, where the original proposal made only registered stock certificates available to foreigners. Senator Othon Mader had proposed an amendment to permit private refineries to increase their capacity without restriction; the original proposal had stipulated that there would be a limit to the capacity of such refineries. Finally, the Senate had proposed permitting the CNP, through Petrobrás, to contract with firms for exploration and exploitation; payment for such operations could be made in the form of a "guarantee of participation in the products of exploitation".[53]

On July 15, 1953, the Chamber of Deputies named a fifteen-man committee and gave it two months to study the Senate's amendments. Most of the committee members were state-monopolists, ranging from the extremism of Euzébio Rocha to the moderation of Maurício Joppert. The only exception was Daniel Faraco (PSD), who had said that financial and technical participation by "the trusts" would be a vitalizing influence on Petrobrás.[54] It came as no surprise that the committee's report, presented September 8, urged rejection of the eleven amendments that would have liberalized the oil law. The committee took particular issue with the four amendments that would have allowed foreigners to buy shares in Petrobrás, even though one had stipulated that foreigners could not buy voting stock. The report heaped scorn upon the Senate for its naïveté in thinking that "the simple suppression of the right to vote would be sufficient to ensure the nationalistic character of the undertaking."

Amendment 28, which would have lifted the restriction on the capacity of private refineries, was branded "one of the propositions that, by indirect and sinuous means, attempts to annul the precepts which institute the state monopoly". And the committee sharply criticized amendment 32, allowing exploration and exploitation by foreign companies under contract. For the committee, this would annul the state monopoly, since it amounted to granting concessions. Furthermore, the amendment was unconstitutional, since it implied the alienation of subsoil wealth.[55]

The Chamber followed the recommendations of its committee in the subsequent voting. On four amendments a roll-call vote was necessary: it showed that the Chamber of Deputies supported a nationalistic oil policy by roughly three to one (support which applied to the three major parties, most strongly to the PTB and least to the UDN).[56] Project 1516/51 was passed September 15 in virtually the same form that had been presented to the Senate a year earlier; President Vargas signed it October 3. After twenty-two months of acrimonious debate the revised Petrobrás project became Law 2004.[57]

In a sense, Vargas got what he originally wanted, a compromise solution to Brazil's oil problem, since both radical nationalists and liberals had denounced Petrobrás. But it was a compromise that was more nationalistic than Vargas wished. Indeed, when one recalls his receptivity to offers from "the trusts" during the Second World War and his concern about allowing investment by foreign capitalists in Petrobrás, Vargas emerges as more concerned with development than with nationalism, quite the reverse of his reputation.

Expert opinion in the United States was unanimous in its reaction: the legislation condemned the oil to remain underground. The *New York Times* wondered if, in the face of Brazil's "chaotic" financial condition (so described, said the newspaper, in a "recent statement" by Finance Minister Oswaldo Aranha), the government would be able "to raise the funds for such a big undertaking as that of producing oil."[58] *The Oil and Gas Journal* and *World Petroleum* said Brazil had simply turned the key in an already closed door.[59]

Most Brazilians, however, saw hope in Petrobrás for the development of national oil reserves without the risks involved in admitting foreign capital. Up to 1953 one of the main problems with the national effort in petroleum had been economic: exploration had been financed entirely by arbitrary government allotments that had never been sufficient. Law 2004, by contrast, set up specific sources of considerable revenue. Indeed, the most severe Brazilian criticism appears to have come from the economist J. Soares Pereira and J. Café Filho (Vice-President of Brazil under Vargas), who criticized the Chamber of Deputies for putting so much of the burden of the company on the federal government. Formerly critical voices were quiet when Petrobrás came into being; they seem to have resigned themselves to "the will of the masses".[60]

It should be noted, however, that the company did have the power to raise revenue to supplement government allotments. For example, it could issue bearer debentures "up to the limit of double its paid-up capital", and could offer stock to Brazilian citizens (preference being guaranteed to government agencies).

Finally, private companies that held refining concessions had to contribute part of their profits to the exploration effort. The planners hoped in these ways to give the new company sufficient impetus to make it self-sustaining so that it would not have to appeal indefinitely to public sources of funds.

But would the financial resources envisaged in Law 2004 be sufficient for the tasks set out for Petrobrás? Deputy Raimundo Padilha had explored the probable costs of setting up an integrated petroleum industry during the debates on the Petroleum Statute. Padilha had calculated that Brazil would have had to spend approximately Cr$13 billion to achieve a production level of 50,000 b.p.d. and create an integrated industry to take care of it. Petrobrás was to start with a capital of Cr$4 billion, which would rise to Cr$10 billion by 1957. Granting that consumption had doubled from 1947 to 1953 (no figures are available for 1953, but the assumption is probably correct, based on 1945–50 figures), the capital allotted would not have been sufficient to set up an integrated, self-sufficient industry, based on Padilha's calculations.

But the same caveat must be raised with the 1953 figures as with Padilha's of 1947: they could be no more than tentative, given the peculiar conditions of Brazil. For example, the cost of exploring in areas remote from the immediate coastline was not known. More to the point, no one knew what oil reserves Brazil really had. The country could pour billions of cruzeiros (and — far more serious — dollars) into an exploration effort that could yield nothing because no one really knew what potential Brazil had in petroleum. Yet the effort had to be made, and Petrobrás was certainly better equipped to do it, from a financial standpoint, than the CNP had been.

But it must be stressed that the effort represented an enormous outlay in foreign exchange for the immediate future. In 1953 Brazil manufactured virtually none of the equipment which would be used in the various phases of the oil industry Petrobrás was to create, and most of the materials would have to be paid for in hard currencies. As we have already seen, Brazil was in the throes of a serious foreign-exchange crisis at that time. The Senate's amendment 32, which would have relieved the drain on exchange in the areas of exploration and exploitation, had been rejected by the Chamber of Deputies because it broke down the state monopoly. Moreover, purchases of refinery and transportation equipment would have to be made in hard currencies for the most part. The adoption of a nationalistic oil policy represented a tremendous burden on the already strained Brazilian economy.

It was certainly public opinion that made Petrobrás a strictly national company, excluding all foreign capital. The Chamber of Deputies was more responsive to public opinion and new forces than the Senate; the former body amended the bill to make it more nationalistic and overrode amendments from the Senate to liberalize the new company. Brazil had changed, had embarked upon a serious program of industrialization; this development provoked the creation of an industrial elite. This new elite was nationalistic, where the old was export-oriented and *inter*nationalistic. In contrast to the *laissez-faire* attitude of the older group, moreover, the new elite favoured state intervention in the economy. It did so because it owed its existence to the post-1930 drive toward industrialization, a drive characterized by active federal-government participation. The passage of the Petrobrás bill in modified form was proof that the new elite had triumphed over the old, and that its program for Brazil was popular.

Petrobrás began formal operations on January 1, 1954, when the new licensing fees on seacraft, aircraft, and motorized vehicles went into effect. No one could get a new licence without proof of a "compulsory contribution" paid to Petrobrás.[61] On April 1, 1954, the CNP (which remained the policy-making organ for petroleum) announced that it had awarded five-year contracts, worth $200 million in all, to the Esso Export Corporation and the California Transport Corporation to supply crude oil for the government refinery at Cubatão, São Paulo, which was expected to begin operations the following year with a capacity of 45,000 b.p.d.[62]

On April 2 President Vargas turned over to Petrobrás all government oil holdings and assets, valued at $165 million. These included the production fields in the Recôncavo (yielding 2,500 b.p.d.); the Mataripe refinery, whose capacity had just been increased to 5,000 b.p.d., and the new refinery at Cubatão;

the national tanker fleet of twenty-two vessels; and drilling equipment scattered throughout eight states.[63]

In its last report before it was relegated solely to a policy-making role, the CNP drew attention to further modest gains. Not only had crude-oil production risen slightly in 1953 to an average of 2,500 b.p.d., but natural-gas production had begun from the Aratú field in the Recôncavo, supplying a cement factory and thermo-electric plant in the region. Mataripe's production had reached 5,000 b.p.d., and studies to treble its capacity had already been completed; plans for an asphalt plant at Mataripe had also been finalized. Finally, the CNL had found indications of oil at Nova Olinda, near Manaus, up the Amazon River.

In its active life as the director of a state-controlled oil industry, the CNP had drilled 404 wells: 367 in Bahia, and the rest scattered throughout Alagoas, Sergipe, Pará, Paraná, Acre, Maranhão, and São Paulo. Of these, 244 yielded oil and 29 gas— all in Bahia.[64] Obviously, it saw its role in exploration as a developmental one, expanding the production capacity of a producing region; it did little wildcatting. The CNP, hamstrung by meagre revenues from government, had guided the national oil industry to a point at which its assets were worth $165 million (however, most of this had fallen to it from the abortive SALTE plan under the Dutra administration). This total was slightly more than three-quarters of Petrobrás's initial capital, all to be subscribed by the federal government. How times had changed!

As first president of Petrobrás, Vargas named Army Colonel Juracy Magalhães, who said in his acceptance speech: "Instead of just talk we are going to work." Magalhães had been a *tenente* and then *Interventor* and later Governor of Bahia; he had contributed nationalistic articles denouncing the Petroleum Statute during the latter half of the following decade. He had just completed a term as Military Attaché in Washington. Geologist and oil pioneer Glycon de Paiva later said Magalhães generated "an atmosphere of confidence and hope that was [Petrobrás's] mystical capital for a long time."[65] Magalhães was pragmatic and positive about foreign assistance. While Law 2004 did not permit Petrobrás to make contracts with foreign firms for the exploitation of Brazilian oilfields, it did not prohibit the employing of such firms for geophysical exploration, test drilling, and the designing of refineries.⁶⁶ Magalhães was thus able to hire known firms for specific tasks.

Vargas appointed three technicians long associated with the Brazilian oil industry as directors of Petrobrás: Irnack Carvalho do Amaral, geophysicist and oil pioneer; João Neiva de Figueiredo, mining engineer with the CNP; and Colonel Arthur Levy, long the representative of the War Ministry on the CNP. On the advice of Irnack Carvalho do Amaral and João Neiva de Figueiredo, Magalhães engaged Walter K. Link, chief exploration geologist for Standard Oil of New Jersey, to staff an exploration department for Petrobrás. Link brought with him excellent men from all over the world and hired the best Brazilian exploration geologists for his department.⁶⁷ It quickly became the largest of its kind in the world, renowned for its quality.

The opponents of Petrobrás, however, had not acknowledged defeat. In a speech to the Higher War College in June 1954, Juarez Távora questioned the wisdom of a state-monopoly policy for national petroleum development. He said he feared that Brazil would be retarded in its quest for self-sufficiency in oil and would be led to "sacrifice unnecessarily, through the exigencies of an exaggerated nationalism, the present rhythm of economic and social development of the country." He called again for the introduction of private capital of any origin, under "rigorous nationalistic control".⁶⁸

Petrobrás and the state oil monopoly were, rightly or wrongly, identified with Getúlio Vargas. It did not matter that he had paid little attention to oil in his first term of office or that he had proposed something less than a state monopoly in 1951; Vargas was the President who had blackened his hand with the oil of Lobato in 1939 and who had signed Petrobrás into existence in 1953.

The strongest opponents of Getúlio Vargas were also the most obdurate opponents of Petrobrás and a state oil monopoly. They included members of the UDN and the "anti-Vargas" wing of the military, led by Generals Canrobert Pereira da Costa and Juarez Távora. Economic liberals, they sought a relatively high degree of foreign participation in national development. During his first term of office Vargas had been able to keep his opponents at bay, but in this second term his judgement was less keen and and his manoeuvring lacked finesse. He had relied again upon the nationalistic wing of the military, but it had collapsed and left him confronted by his old army enemies. A series of unfortunate economic moves had alienated almost every economic sector, including his traditional supporters, the masses. So in August 1954, when one of Vargas's aides engineered an assassination attempt on Carlos Lacerda, an implacable enemy of Vargas, the President found himself isolated and vulnerable. On August 23, twenty-seven army generals called for his resignation.

But the President confounded his enemies once more: he committed suicide the next day. An inflammatory suicide note he was said to have written charged that "a subterranean campaign of international groups joined with national groups" had conspired against him in his efforts to help Brazil, and had tried to block such proposals as Petrobrás. The note closed: "I offer my life in the holocaust. I choose this means to be with you always." One commentator termed this note "the strongest national appeal [Vargas] had ever made".[69] At any rate, it rendered opposition to Petrobrás impotent for years. By his final act, too, Vargas tied his name to Brazilian petroleum in the national conscience. He became identified with the great gamble: could Brazil achieve her goal of self-sufficiency in oil? Should she ultimately reach that goal and gain "economic emancipation" (in Vargas's words), his fame would be all the greater.

5
YEARS OF SUCCESS, 1954–1960

Petrobrás, despite its politicized and long-delayed birth, began operations auspiciously under policy guidelines that the CNP laid down. The CNP decided to use private enterprise as fully as possible where Law 2004 permitted, and refused to extend Petrobrás's monopoly beyond areas which the law specifically described.

For example, early in 1954 the CNP called for bids to supply Brazil's crude-oil needs for the next five years. While some criticized this as a tacit admission on the part of the CNP that Petrobrás would not contribute significantly to consumption in that period despite the enormous revenues to be allotted to exploration, the CNP decision was both reasonable and beneficial to the exploration effort.[1] An oilfield normally takes several years to develop before its production can reach the market; five years does not seem too long. Further, DEPEX (as the Exploration Department was known) would be able to concentrate on its role — exploration — without the added pressure of meeting consumption requirements. Fields could be developed properly, without undue haste, and excess production could be sold abroad. Indeed, such excess production would have to be sold elsewhere, since only the Mataripe refinery (with a capacity of 5,000 b.p.d.) was equipped to handle Brazil's high-paraffin crude; private refineries could handle only foreign crude of certain types, and Petrobrás's almost-ready Cubatão refinery had been built to process Venezuelan oil.

The CNP also turned its attention almost immediately to the petrochemicals industry, and made it quite clear that it favoured private development for it. The stated reason was the "great diversification" that characterized the industry. Here again, Petrobrás was being relieved of a potential distraction, and a costly and bewilderingly complex one at that. It is clear that the CNP saw Petrobrás's main function as finding enough oil in Brazil to satisfy consumption requirements, and at the same time saving precious hard currencies by becoming self-sufficient in refining and transport, so that Brazil would import only the cheapest commodity — crude oil — in her own tankers.

Both the CNP and Petrobrás sought to use as much foreign technical and industrial assistance as possible, until such time as Brazil could hope to supply both. By early 1955 Petrobrás had signed contracts worth $300 million with foreign concerns; most were for crude-oil supplies, but some were for refinery construction, exploration, technical assistance, and the purchase of tankers.[2]

The most important foreign expert was Walter Link, whom Petrobrás hired to form DEPEX. Link took a highly practical view of his task, declaring after his initial tour of sedimentary basins that "since Petrobrás is in its infancy, struggling to get on its feet and become self-sufficient, we must explore the regions that offer the best prospect and are of easiest access."[3]

Petrobrás's organization facilitated such a pragmatic and businesslike attitude. It was, like the great international companies and such state entities as Mexico's Pemex, a vertically integrated holding company. Considerable autonomy for each segment of the company was set down initially, and the administrative integration occurred at the executive level only. Theoretically at least, then, refining, transport, exploration, production, and other such sectors could get on with their own tasks without pressure from the others. That is, as long as the calibre of administrators was high and all sectors developed at approximately the same rate.[4]

Colonel Arthur Levy, a charter director of Petrobrás, became president of the company late in 1954, after João Café Filho succeeded to the presidency of Brazil for the remainder of Vargas's term (almost a year and a half). Levy, who had supervised the construction of the Santos–São Paulo pipeline and the refinery at Cubatão, continued the work of his predecessor, Juracy Magalhães, in an equally vigorous spirit. He ran the company like a true corporation president, making decisions in conference with his highest executives. Morale was high.[5]

Levy's appointment set two trends for the selection of Petrobrás's chief executive which have persisted to the present. First, Levy was a career army officer as Magalhães had been; there have been only two civilians appointed to the office since, despite a high turnover. This leads to the second trend, which was that the chief executive of the company changed with that of the country. Under normal circumstances a five-year term as president of Petrobrás (coinciding with the constitutional term of Brazil's president) would have been reasonable for the individual and the company. Brazil, however, had five presidents in the first decade of Petrobrás's existence,* and the company even more chief executives. Consequently, there was little continuity of policy at the top level in Petrobrás.

After Vargas's suicide public debate over oil policy continued, although on a much more moderate scale, and traditional opponents of a nationalistic solution persisted in their attacks upon Petrobrás. They were encouraged by the fact that Café Filho was more receptive than Vargas had been to the participation of foreign capital in the economy. (Café Filho's Cabinet also reflected a "liberal" outlook, especially with the appointment as Finance Minister of Eugênio Gudin, who had held many positions with foreign firms operating in Brazil and whom radical nationalists regarded as an arch-*entreguista*.)[6] For example, in September 1954 the Federation of Commercial Associations, an unflinching opponent of Petrobrás since 1951, called on the new government to allow foreign capital to help develop half the oil lands, reserving the remainder to the state. Otherwise, the federation believed, a proper survey of Brazil's oil potential would never be made.[7]

The Rio newspaper *Correio da Manhã* also continued its implacable opposition to a "state-monopoly" solution. In October 1954 it condemned the "narrow policy of slogans, the . . . demagogic nationalism" which had brought about the creation of Petrobrás, and charged that Brazil's oil had been condemned to

---

* The Brazilian presidents were: Vargas (1950–54), Café Filho (1954–55), Kubitschek (1955–60), Quadros (1960–61), and Goulart (1961–64).

remain underground, "unexploited and useless, for lack of resources". The daily considered that the "crisis in petroleum, as in other problems, is a consequence of economic subjects having been dealt with in a spirit of petty electoral and demagogic politics."[8]

Senator and newspaper baron Assis Chateaubriand continued his opposition to Petrobrás, repeatedly holding the nationalistic oil policy up to ridicule, or condemning it as communistic, in the Senate and in his newspapers.[9] One such attack, late in October 1954, prompted the newspaper *Diário de Notícias* (formerly an organ of the Petroleum Centre) to publish a series of articles calling upon all Brazilians to unite behind Petrobrás to give it a chance to prove itself. Messages of support poured into the paper.[10]

Late in October 1954, however, Juarez Távora, who had fought against Petrobrás and restrictive nationalism, declared that he had changed his policy. The general reaffirmed his faith in a "mixed" solution, but, since the government had chosen a monopolistic policy, he called upon his countrymen "to close ranks, patriotically, around this solution, according it . . . the support that it should have". Eventually, he hoped, the public would come to support his preference and induce the government to adopt a more liberal oil policy.[11]

Távora's "defection" indicated that the military had closed ranks and were now prepared to support the state company, at least for a time. Távora, incidentally, had by then emerged as a probable candidate for the Brazilian presidency; his policy change may have been in response to strong public support for Petrobrás. Indeed, with the exception of Plínio Salgado, ex-leader of the fascistic *Integralistas*, all the candidates in the 1955 presidential election firmly endorsed Petrobrás (this included even Etelvino Lins, for a few months the candidate of liberal factions of the PSD and UDN).[12]

Finance Minister Eugênio Gudin, seeking to put the Brazilian economy on what he saw as a sounder basis, tried on the one hand to restrict credit and on the other to change Law 2004 so as to open the oil industry to foreign-capital participation. In the former pursuit he antagonized the business community, and in the latter he alienated such important figures as Juarez

Távora and President Café Filho (who later revealed that he had refused an offer from Standard Oil of New Jersey to invest a substantial sum in exploration if suitable changes in Law 2004 were made). Moreover, the armed forces, through General Canrobert Pereira da Costa, declared that it would be "inopportune" at that time to alter the state monopoly on petroleum, without a trial period for Petrobrás.[13]

In March 1955 Petrobrás brought in a producing well at Nova Olinda, on the Madeira River near Manaus, stirring hopes of a second petroleum zone outside the Recôncavo, and a new boom for the Amazon region.[14] Drilling had begun at Nova Olinda in 1951 under the CNP, and oil was found about 9,000 feet down. The region was about fifty miles from Manaus and 1,000 miles from the mouth of the Amazon River. The well was on a riverbank, which facilitated shipment of the oil produced. But further exploration of the zone would be difficult because of the jungle.[15]

Shortly after this important discovery, geologist Glycon de Paiva published an article discussing Nova Olinda. Prior to the discovery, findings in the area had not led to any optimism about its oil potential. The sediments had appeared to be "rather young and not very thick", and consultants of Geophysical Services, Incorporated, had expressed "disinterest" in drilling. Dr. Décio Savério Otoni, a geophysicist and chief of the Exploration Service of the Amazon had, however, insisted on drilling an exploration well. Glycon de Paiva wrote, "Possibly he discerned another interpretation among those which the twenty-odd American geophysical engineers gave to the case." Otoni converted Plínio Cantanhede, director of the Amazon Region, to his position, and the Nova Olinda "gusher" was the result.

The discovery, concluded Glycon de Paiva, indicated that there was indeed oil in commercial quantities within a structure of some breadth and depth at Nova Olinda. This was of even greater significance when one considered the geologic uniformity of the Amazon basin. The geologist concluded that it would take five years to discover the commercial potential of the Nova Olinda region.[16]

At this time (early 1955), Brazil's search for oil suddenly and dramatically became international, when Bolivian President

Víctor Paz Estenssoro complained bitterly to his Brazilian counterpart, João Café Filho, about Brazil's apparent disinterest in prospecting in the area reserved to it under a treaty signed by the two countries in 1938 (see Chapter 2). Bolivia's state oil company, Yacimientos Petrolíferos Fiscales Bolivianos (YPFB), had brought in a major discovery at Camirí near the treaty zone early in 1954, and President Paz was anxious to extend the exploration area. Bolivian nationalists—the most ardent supporters of Paz's MNR government which had come to power by revolution in 1952—saw Brazil as imperialistic because of her reluctance to explore the region, and demanded immediate action from Brazil or threatened to break the treaty. Ironically, this was the same accusation Brazilian nationalists had made against "the trusts" twenty years earlier.

Café Filho managed to have a decision postponed until after the forthcoming presidential election in Brazil. His Finance Minister, Gudin, had informed him that it would be "difficult, if not impossible" to find the $4 million needed to comply with the treaty, and Café Filho, in the last months of his short term, wanted simply to finish with as little controversy as possible. Bolivia could be kept waiting.[17]

Within Brazil the fact that Petrobrás was beginning to show results and was being well run did not stop challenges to it from within the country. Just after the Nova Olinda discovery, for example, the Senate voted down another proposal to permit participation of foreign capital in developing oil resources.[18] A year later another senator sought in vain to restrict Petrobrás to a radius of twenty-two kilometres from the pioneer well at Lobato and the new well at Nova Olinda. The remainder of the country would be opened to exploration and exploitation by concession.[19] Undoubtedly, the determining factor in Petrobrás's strength was the nationalism strengthened by the suicide of Vargas, which had forced such liberals as Juarez Távora to change his policy and had also ensured widespread military support for the company. Indeed, Vargas's suicide had prevented the Café Filho administration—particularly such internationalists as Eugênio Gudin—from diluting any aspect of the economic nationalism the late President had fostered. The change of administration had had no effect upon the nationalistic course upon which Brazil had embarked under Vargas.

At the other extreme, a proposal was made in the Chamber of Deputies in 1956 to bring the distribution of petroleum derivatives under the state monopoly.[20] However, Colonel Janari Nunes, whom Juscelino Kubitschek had made president of Petrobrás in 1956, told the Chamber of Deputies that the company "at the moment" did not want the responsibility of a monopoly on distribution.[21]

The reasons for this proposal must have been entirely political: The CNP held sole authority to set product prices, and the profit margin of distribution — despite radical rhetoric to the contrary — was small. The cost of nationalizing the distribution network would have been staggering, and purposeless from an economic standpoint. It made sense only in the realm of populistic politics, where "completion of the state monopoly" (a cry which would be heard with increasing frequency in the early 1960s) sounded logical.

A new weekly tabloid, *O Semanário*, took up the task of defending Petrobrás against any and all criticism and of urging complete state monoploy. It issued frequent statements about the depth of mass support for Petrobrás, and periodically warned the public about the constant threat "the trusts" posed to Petrobrás (which the paper always characterized as the key to the economic emancipation of Brazil). For example, *O Semanário* carried an exclusive interview with Plínio Cantanhede, former president of the CNP and then director of the Amazon region for Petrobrás; Cantanhede declared that the force of public opinion behind Petrobrás compared with that in favour of the abolition of slavery in Brazil.[22] The journalist Joel Silveira, a frequent contributor to *O Semanário*, declared in a 1956 issue that "the trusts" were reaping the benefits of all the hard work done by Petrobrás because they controlled the petrochemicals industry, which was the most profitable part of the petroleum enterprise (said Silveira).[23] The weekly apparently took over from the defunct *Jornal de Debates*, which had flourished in the late 1940s, but it was considerably more sensational and hysterical in tone than its predecessor.

Late in 1955 Bolivia again became of concern to Brazilian policy-makers, when President Paz Estenssoro, short of capital and foreign exchange and anxious to find a viable export commodity, decreed a petroleum code which opened up almost all national territory to exploitation by concession. Brazil, also short of foreign exchange and under pressure from Bolivia to explore the treaty zone or abandon its rights there, might well have done the latter had not influential military and civilian defence experts counselled against such a move. General Canrobert Pereira da Costa, Chief of Staff of the Armed Forces, the Chiefs of Staff of the Army and the Navy, and the National Security Council all recommended implementing the treaty in the hope that Brazil would find a continental source of oil which could be tapped by land transport (rail or pipeline). Then Brazil would not need to rely upon overseas sources, unreliable in wartime.

So, before leaving office, President Café Filho instructed his Foreign Office to initiate discussions with Bolivia. By December 1956 the Foreign Office was interviewing Brazilian companies which might be authorized to develop oil on Brazil's behalf in the Bolivian treaty zone. As talks progressed, Bolivia made clear its opposition to Petrobrás's participation in the discussions and in the exploitation of the treaty zone; only private companies were to develop the zone under the terms of a prospective agreement.

Unfortunately for the negotiators, Colonel Janari Nunes, president of Petrobrás, insisted on his right to participate in the discussions. From a Brazilian standpoint he was entitled to do so because of Petrobras's pivotal role in Brazil's oil industry. Although Law 2004 gave Petrobrás a monopoly over pipeline and maritime transport of oil, Bolivian oil might have been exported to Santos via the just-completed railway from within Bolivia to Brazilian private refineries or the outside world; Petrobrás had not been given control of rail transport of oil.

Bolivia's negotiators gave no reason for their intransigence over Petrobrás, and one is left to speculate. A possible factor was mooted in a liberal Rio newspaper: Petrobrás had been founded upon anti-imperialism and radical nationalism, and it was an embodiment of the state; for it to have operated in Bolivia would have been nothing short of an invasion of a sovereign nation. On closer examination, however, this argument does not make much sense; indeed, it could just as easily have been used to *justify* Petrobrás's exploration and development of the treaty zone. At this time Italy's state oil company, Ente Nazionale Idrocarburi, was skilfully using nationalistic rhetoric to bid against "the trusts" for exploration rights in other countries! True, crowds denouncing "Brazilian imperialism" had greeted Café Filho in Bolivia during his visit in January 1955, but the charge at that time arose out of Brazil's inactivity in the treaty zone; at no time was such a charge used to forestall Petrobrás's participation in development of the treaty zone.

There is another possible reason for Bolivia's attitude: it wanted to have the treaty broken so that it could open the zone to exploration and exploitation by "the trusts". Under the 1938 treaty, Brazil was exempted from exploration taxes; foreign private companies not only had to pay such taxes but pay them in dollars, and Bolivia was desperately short of foreign exchange. The Bolivian subsidiary of Gulf Oil had expressed considerable interest in the treaty zone, and Bolivian negotiators were apparently putting pressure on their Brazilian counterparts to abandon the treaty.

Negotiations over the treaty continued in a desultory fashion until the last months of 1957. Itamarati (the Brazilian Foreign Office) continued to insist on Petrobrás's right to participate in the deliberations, while at the same time considering proposals from Brazilian private companies for exploring and developing the treaty zone. Bolivia continued to oppose Petrobrás's participation.[24]

Meanwhile, Brazil was beginning to "nationalize" again — if rhetoric and newspaper attitudes are an indicator — after the lull that had followed Vargas's suicide in 1954. The chief stimulus was Juscelino Kubitschek, a skilled politician from Minas Gerais who took office as President of Brazil in 1956. Kubitschek won the election with much the same support as Vargas had had in the 1950 election: old-style political "bosses" and urban labour — the latter through his alliance with João Goulart, vice-presidential candidate and heir to Vargas's PTB support. Immediately upon his election, Kubitschek initiated a program to galvanize Brazil into development; its focus was the building of Brasília deep in the interior, a move to give Brazilians new pride. The means of development was to be import-substitution, with massive amounts of the state's resources poured into key bottlenecks in the economy. One of the bottlenecks was energy, of which petroleum was a vital part.[25]

Thus, it was not long before Petrobrás was in the news again, as a symbol of the "developmentalist nationalism" Kubitschek was promoting as well as of the persistent undercurrent of radical nationalism which had created the company (see Chapter 3). Perhaps the catalyst that prompted Kubitschek to begin using Petrobrás as a symbol was a warning, published in *O Semanário* in May 1957, from the War Minister, General Henrique Teixeira Lott. He was quoted as saying to Kubitschek, "I regret having to say to your Excellency that, should the Petrobrás law be modified, the Army will find itself wondering whether to withdraw completely its support from the government."[26]

A few months later, Lott told a student delegation that the only way Brazil could "emancipate" itself was through the nationalization of its mineral resources; "most" of the armed forces were of this opinion, he added. (This statement apparently aroused some controversy, because three days later the newspaper that had reported the remark defended the right of the military to voice opinions on subjects of national moment.)[27] Lott had first come to national prominence while War Minister in Café Filho's caretaker government, as a key figure in the controversy over whether Goulart and Kubitschek would be allowed to take office, since they were closely identified with the *"getulistas"*. The military was split, with one group, identified with Carlos Lacerda and part of the UDN, calling for a military *coup* to block the return of the *getulistas*, and the other, with which Lott became allied, advocating a strictly legalistic course and allowing the new administration to take office. Lott's decision to abide by the constitution, and his subsequent moves to uphold the course of action dictated by it, thrust him into the limelight and drew him inexorably into the nationalistic camp, which was dominated by the *getulistas*. By mid-1958 Lott had become a champion of Brazilian nationalism, and was clearly looking to the presidential elections of 1960. In August 1958 the Nationalist Parliamentary Front (FPN), a loose coalition of nationalistic deputies from several parties, which wielded considerable power until the 1964 revolt, pledged its support for Lott and his nationalistic stand, particularly on Petrobrás.[28]

Lott's remarks are significant on two counts. First, he apparently represented the more nationalistic elements within the officer class, who were enjoying a resurgence of influence (their representatives won the 1958 Clube Militar elections). Lott thus spoke for a group crucial to Kubitschek's freedom of action, or even his survival. Second, Kubitschek seems to have decided some time early in 1957 that although he was offering something for everyone in his developmentalist nationalism, he would cater more openly to nationalists in the hope of securing a mass political base. In the process he would alienate the same groups that had opposed the formation of Petrobrás — the traditional, export-oriented business elements and their allies — but it is doubtful that they had ever supported him. Thus, by mid-1957, Kubitschek had begun to charge that there was a campaign against him that had, as one of its chief objectives, "the weakening of the nationalistic oil policy". The President accompanied this charge with declarations of support for Petrobrás and state industrialization.[29] Clearly, Petrobrás was to become a political pawn again.

The year 1957 marked the beginning of a tradition: the use of certain dates for nationalistic rallies. For example, on April 19 of that year (the anniversary of Vargas's birthday) the Clube Militar heard an address by Petrobrás president Colonel Janari Nunes, in which he charged that the international oil companies encouraged dictatorial regimes in order to maintain their grip on reserves and hold wages to low levels. He said Brazil was fortunate in being one of the few countries in the world outside the orbit of "the trusts". A member of the audience who expressed doubt that Brazil could be self-sufficient in oil was roundly booed.[30] At a patriotic rally on May 1, 1957, Vice-President Goulart pledged undying support for Petrobrás, which he asserted was "today completely victorious".[31] Kubitschek in turn condemned "forces which conspire against the economic development of Brazil". He did not name them, but *O Semanário* said darkly, "We all know what forces those are."[32]

Independence Day (September 7) also became an occasion to glorify nationalism and Petrobrás. Also, each year on August 24, the anniversary of Vargas's suicide, nationalistic rallies, using Petrobrás as a symbol, were held in Rio de Janeiro's Cinelândia plaza (where a bust of Vargas stood). The radical-nationalist daily *Última Hora* used the latter occasion to reprint the ex-President's suicide letter, with a picture of Vargas holding out his hand smeared with oil. In the 1958 anniversary issue, for example, Horta Barbosa declared, "To overthrow Petrobrás would be to betray Brazil."[33]

Petrobrás was clearly becoming the focal point in the bitter controversy over Brazil's development policy. As the epitome of nationalism, it was either the point of attack by "liberal" forces or was thrust into the front lines of the "battle for economic emancipation" by the nationalists—developmentalist or radical — as an object of uncritical, emotional praise. Not a healthy atmosphere in which to develop a national oil industry.

In August 1957 Janari Nunes revealed that the company had found oil in Alagoas state. Although the announcement gave no further details it was greeted with celebrations in Maceió, capital of the state. Students were granted a "Petroleum Week", and the Archbishop of Maceió affirmed the support of the Church for Petrobrás. For Petrobrás was indeed yielding impressive results (even though production still came only from the Bahia Recôncavo), and Brazilians had reason to be proud and optimistic. The production goal of 40,000 b.p.d. that Kubitschek had set for 1960 had been passed by the end of 1957. Now President Kubitschek said he wanted 100,000 b.p.d. by 1960 (output had been 6,500 b.p.d. when he took office). Of course, at the same time consumption was rising; for 1957, the increase in production was only enough to ensure that the dollar amount spent on foreign oil would decline slightly over the 1956 figures.[34] Still, production *had* kept pace with consumption!

Perhaps the most important task for Petrobrás — apart from exploration — was to make Brazil self-sufficient in refining as quickly as possible (to substitute Brazilian products for imported, in the long run saving precious foreign exchange). By the end of 1957 the refining picture was showing gains at least as impressive as those in production.

In 1954 two privately owned refineries had at last begun production near Rio de Janeiro and São Paulo. The former, at

Manguinhos, a suburb of Rio, was owned by Refinaria de Petróleo do Distrito Federal. It had a capacity of 10,000 b.p.d., and was to use Venezuelan crude oil supplied by Standard Oil of New Jersey. The latter, at Capuava, between São Paulo and Santos, was owned by Refinaria e Exploração de Petróleo União. Its capacity was 20,000 b.p.d. (raised to 31,000 by the CNP in 1957) using crude from Saudi Arabia supplied by Gulf. Early in 1957 President Kubitschek opened a refinery with a capacity of 5,000 b.p.d. at Manaus, on the Amazon River. Crude oil for the plant was to come down the Amazon from the Ganso Azul field in Peru. The Companhia de Petróleo da Amazônia owned the refinery, and it had been built by the Southwestern Engineering Company, of Los Angeles (see Chapter 3 for the planning of these refineries). These three joined the existing Ypiranga refinery, with a capacity of 9,300 b.p.d., and the two other tiny plants in São Paulo and Rio Grande do Sul.[35]

Petrobrás was also carrying out its charge in the refining sector (by Law 2004, no more private refineries could be built, nor could the existing private refineries increase their capacity). In April 1955 the state company inaugurated a 45,000 b.p.d. refinery at Cubatão, a new industrial centre between São Paulo and Santos. The plant was named after Artur Bernardes (who had died a month earlier), ex-President of Brazil (1922–26), and a vigorous champion of Brazilian nationalism throughout his life (ironically, the refinery was designed to process foreign crude). By 1958, capacity of the refinery had been increased to 70,000 b.p.d., and a petrochemicals plant added to the site. Petrobrás had also announced plans to build a 90,000 b.p.d. refinery at Duque de Caxias, in the state of Rio de Janeiro, on Guanabara Bay close to the city of Rio.[36]

An enterprising reporter for the *New York Times* discovered that Petrobrás was relying heavily on "the trusts" for technical assistance and financing in the building of her refineries at this time, something of which the Brazilian populace was unaware. For example, the Texas Company and Standard Oil of California were helping to finance Duque de Caxias refinery, and the Foster Wheeler Corporation had been awarded a contract for $40 million to build it. According to the same source, "an American company" had helped finance the terminal and oil docks in Santos for the refinery at Cubatão, and Standard Oil of California had lent Petrobrás $1 million for an asphalt plant. Petrobrás had also secured financing for various projects in Europe; but although European sources offered better terms, being anxious to crack the Latin-American market, Brazil was already heavily committed to American equipment and technology.

This is not to say that such financing was illegal; quite the contrary. Law 2004 did not prohibit it, and the CNP and the original executive of Petrobrás were convinced such assistance was vital for the rapid growth of the company. However, in the political climate which produced Petrobrás and which continued to "guard" it, to have mentioned how closely Petrobrás was working with "the trusts" would have been a gaffe of substantial proportions. Better to draw attention—as Petrobrás did—to the long-term financing the Italian *government* extended to the company to purchase drilling and refinery equipment, and to the technical assistance Ente Nazionale Idrocarburi, the Italian state oil company, was supplying it.[37]

Despite the generally optimistic picture presented in the refining and production sectors, some clouds were gathering. In October 1957 the Chateaubriand newspaper *O Jornal* had questioned the use Janari Nunes was making of Petrobrás public-relations funds. The newspaper charged that not only had allocations for publicity substantially increased since Nunes had taken office, but that publicity now had "for its exclusive end" the promotion of the "political aspirations of its President". The newspaper declared that Nunes wanted to be "carried to the Presidency of the Republic on the shoulders of the nationalists".[38]

The Petrobrás image was further compromised in April 1958, when the administrative competence of Nunes was publicly called into question. A letter written by João Neiva de Figueiredo (one of two Petrobrás directors who had resigned in November 1957), charging Nunes with gross mismanagement, had fallen into the hands of the pro-Petrobrás newspaper *Diário de Notícias*. The newspaper had asked permission to publish the letter, but Figueiredo refused lest such a move jeopardize the prestige of Petrobrás. The paper investigated Figueiredo's charges, found them to be true, and decided to publish the letter anyway, to show what Nunes was doing to the company.[39]

In his letter Figueiredo complained that Nunes could not work with a group, and was acting without consulting his directors; the "pressures of politics or publicity" were affecting the location of the "undertakings" of Petrobrás, and Nunes was going to the public with declarations about projects "which had not been studied sufficiently". Priorities for investment had not been established with economy in mind, said Figueiredo. Nunes seemed to feel money was to spend or even to waste, and Petrobrás could not afford that. Finally Figueiredo said he considered that Petrobrás was placing too much emphasis on drilling, without "the preliminary and indispensable work of geology and geophysics". As a result of all these factors, he and fellow director Irnack Carvalho do Amaral both resigned.[40]

In a campaign to have Nunes removed, the *Diário de Notícias* examined each charge in detail. Nunes, it reported, had reduced prices of derivatives, exaggerating profits to justify the move. The company needed all the revenue it could get, asserted the paper; Nunes had reduced prices for political motives.[41] The newspaper further charged that he had accelerated crude production without building storage facilities to handle it; in the end he had had to use tankers for the purpose. Nunes admitted this but said he had worked out an agreement with foreign companies to exchange Brazilian oil for crude that national refineries could handle.[42]

Another complaint centred on proposals to build a refinery in Minas Gerais. The refinery would require a pipeline from Guanabara Bay for crude supplies, and to tackle both projects meant that Petrobrás would have to defer several other, and perhaps more important, undertakings — notably expansion of exploration. Several senior Petrobrás officials were against the plan, but no decision had been taken when Janari Nunes told an audience in Belo Horizonte (the capital of Minas Gerais) that their state had a "right" to a refinery. Apparently the CNP remonstrated with Nunes over his conduct, and he responded in a letter which the CNP returned as "insulting".[43] The controversy continued for about a month longer, during which time the campaign waged by the *Diário de Notícias* degenerated into personal attacks on Nunes as it declined in intensity.[44] For the first time, public confidence in Petrobrás had been shaken by a scandal.

Nunes remained in office but the affair was not ended. By November he and Colonel Alexínio Bittencourt, president of the CNP, were involved in a dispute so bitter that Kubitschek had to form an investigating committee. Bittencourt charged that Petrobrás had made many policy decisions, such as acquiring seven "supertankers", without consulting the CNP. He also complained that Nunes had wasted Petrobrás's resources by overinvesting in new refineries and ships. The foreign-exchange situation was then critical, due to a steady decline in coffee prices and a high rate of imports and profit remittances by foreign-owned companies. The dollar value of the cruzeiro had been halved over the previous twelve months.[45]

Última Hora predictably condemned Bittencourt for undermining confidence in the state oil company. One of the paper's columnists charged that the CNP president had been "selected by the entreguista forces to lay siege to Petrobrás in a drastic and mortal manner": it was all part, the columnist charged, of the "demoralization" technique employed by "the trusts".[46]

The controversy was resolved only when, on December 8, Nunes and Bittencourt resigned. Air Force General Henrique Fleiuss became president of the CNP, and Army Colonel Idálio Sardenberg, the choice of War Minister Lott, took over at Petrobrás.[47] The squabble brought warnings from two commentators about the danger involved in surrounding the company with emotion. Hélio Beltrão, a former Petrobrás director who had been responsible for its organization in the image of a private company, appealed to Brazilians not to drag Petrobrás down into the realm of party politics and "ideological conflicts". The company had been "born under the happy sign of multi-party consensus", and should not be tied to the fortunes of any political group.[48] Hélio Jaguaribe, a noted social scientist, similarly lamented the "mysticism" that clouded the Brazilian petroleum picture. At the same time he admitted that Petrobrás could not have been created without it, nor would the company long survive its dissipation. He hoped, however, that the CNP, the policy-making body, would become the focus for emotion, leaving Petrobrás to get on with the serious business of finding oil.[49]

But Petrobrás remained the focus of emotion because it—not the CNP—was discovering Brazilian oil. This fact alone made the state oil company the greatest single factor in the "economic emancipation" of Brazil. The CNP may have been the policy-making entity in theory, but Petrobrás made the policy come to life; indeed, under Janari Nunes it had taken the initiative in policy formulation.

Meanwhile, other events at home and abroad were affecting the picture. The stalemated negotiations in Bolivia had become a subject for debate in the press of both countries by the end of 1957. Some of the more nationalistic Brazilian newspapers were charging that Bolivia was "under the control of the trusts", and their Bolivian counterparts were responding by charging Brazil with "imperialism". Indeed, Bolivian newspapers went so far as to accuse Brazil of stirring up separatism in the southern province of Santa Cruz de la Sierra, in order to annex it and its new-found oil.[50]

Finally, in January 1958 the two sides decided to make a last, determined effort to resolve the dispute, and negotiating teams met at Roboré, a town on the railroad Brazil had built in Bolivia. Clearly, Brazil could hope to stay in Bolivia only if it reduced its claims, and the army was still anxious that it acquire a continental source of oil against the eventuality that Petrobrás might not attain self-sufficiency in Brazilian production. On the other hand, Bolivia wanted an outlet through which it could export its oil and other commodities, and Brazil appeared to be the most friendly neighbour at the time. On January 27 both parties reached an agreement, the "Roboré Accord".[51]

The sections of the agreement that dealt with the 1938 oil treaty reduced the area open to Brazilian exploitation from 34,000 to 13,500 square kilometres, and limited its development to "private companies capitalized in Brazil", subject to the Bolivian Petroleum Code. Brazil agreed to purchase, with hard currencies at current world prices, 100,000 barrels a day of crude oil produced in Bolivia (by any company), 5,000 of which were to be supplied by YPFB, the Bolivian state oil company. The private Brazilian firms were to exploit the Bolivian field solely for Brazilian consumption; if their production exceeded 100,000 b.p.d., they were free to dispose of the surplus as they wished. The oil produced was to be given free passage throughout Brazil and be subject only to normal tariffs.[52] How the oil was to be transported was not resolved: Bolivia favoured having Brazil build a pipeline, but this would have been prohibitively expensive for Petrobrás (the only company allowed to build pipelines in Brazil, by Law 2004); Brazil favoured using the railway.

Brazil's nationalistic press was initially enthusiastic about the Roboré Accord. Rio's *Última Hora*, for example, hailed the pact as "one of the greatest negotiations in our history" and said that Brazil had been lucky to get so much, since it had been dealing not only with Bolivia but with "the trusts", who controlled it and were determined to destroy Petrobrás. The newspaper saw the agreement as a "victory for the realistic nationalism" of President Kubitschek, and heaped praise upon Foreign Minister Macedo Soares, who had headed the Brazilian negotiating team (he had also participated in talks leading to the 1938 treaty).[53]

The National Bank of Economic Development (BNDE) accepted proposals until September 12, 1958, from groups interested in exploiting the treaty zone. By the deadline, six had applied (one of which was a subsidiary of Refinaria e Exploração de Petróleo União). According to the *New York Times*, most of them were negotiating with U.S. and West European interests for the financing of their operations.[54] The BNDE, while considering the applications, indicated that it would favour those from groups which had their own sources of funds. The BNDE did not have the dollars to finance such a venture.

For *Última Hora* such a preference by the BNDE represented a tacit acceptance that "the trusts" would help finance the Brazilian companies developing Bolivian oil for Brazilian consumption. Such companies would be fronts by which the international companies could penetrate Brazil and break Petrobrás and the state monopoly on oil; the newspaper charged that there were such figurehead organizations among the six applicants.[55] The CNP then took up the fear, officially drawing it to Kubitschek's attention. The BNDE, whose president, Roberto de Oliveira Campos, was a liberal economist who had already been a target for the radical-nationalistic press as an *entreguista*, delayed for several months its decision on which companies should go to Bolivia; the controversy over how Brazil should develop the treaty zone continued unabated.[56]

What made the controversy more serious than it might have

been was that negotiations over financing coincided with a very unpopular anti-inflation drive by the Brazilian government. In response to rapidly growing inflation and foreign indebtedness, as well as strong pressure from the International Monetary Fund to apply a "shock treatment" approach to stabilization, Kubitschek attempted in 1958–59 to introduce a measure of austerity into parts of the national economy. Such a move predictably stirred widespread resentment, particularly from nationalists (developmentalist as well as radical), against "foreign pressures".[57] In such a climate, foreign financing for companies proposing to exploit the treaty zone in Bolivia would be bound to arouse strong reaction from nationalists as another example of foreign pressure.

Despite the rhetoric of radical nationalists, however, the BNDE approved in January 1959 the applications of three companies to explore in Bolivia. Under questioning by a Chamber of Deputies committee, all three revealed that they had made arrangements for some financing with U.S. companies. In March the CNP recommended that the three companies be allowed to commence operations, provided they could prove financial capacity to the satisfaction of both governments.[58]

Thus the Kubitschek administration had become "liberal" — at least over this matter — after having apparently turned chauvinistic no more than two years earlier. Radical nationalists railed in vain: cries that Bolivian oil would become a "torpedo against Petrobrás" fell on apparently deaf ears, and demands from nationalistic Congressmen that the Roboré Accord be submitted to the legislature as a new treaty were blithely turned aside by Itamarati's assertion that the agreement was merely "enabling legislation".[59]

There were two reasons for the switch in thinking (if it was that, rather than just an attempt to deal minimally with a touchy problem). First and foremost, comparatively little political capital could be gained from finding oil in Bolivia, since Petrobrás was not involved in the effort; much, on the other hand, would accrue from finding oil in Brazil (which Petrobrás was doing). Second, the austerity program, in the face of inflation and foreign indebtedness, forestalled any possibility of government supplies of foreign exchange finding their way into the Bolivian venture. Precious dollars had to be allotted to the industry that Petrobrás was being called upon to create. Still, the approved companies had received an official go-ahead and were supposed to have adequate financial resources for their ventures. Yet almost nothing came of the search in subsequent years, and the reasons were the same as before, political and financial.

The remainder of the story is depressingly easy to tell. Two companies encountered financial difficulties and folded before even getting to Bolivia. A third (added in the early 1960s) withdrew after some exploration and a tussle with the Bolivian government over alleged smuggling. The fourth did some drilling, but withdrew for lack of funds. The Brazilian government was less than helpful, refusing to allow this last company (the subsidiary of Refinaria e Exploração de Petróleo União) to go to its shareholders for more capital. There the matter has rested.[60]

It is hard to avoid the conclusion that the four companies had about as clear an idea of the amount of money their task would require as the various companies formed to explore in Brazil during the 1920s and 1930s did. There seems to have been a similarly naïve belief that the area was choked with oil, and that the job of extracting it would be simple—and inexpensive—in the extreme. Finally, the official indifference or even hostility to the venture confirms that there was no alternative to Petrobrás in the contemporary political situation.

To turn back to Brazil and 1958, radical nationalists lost an ally when the Brazilian Communist Party (PCB), following international Communist policy, suddenly switched from staunch support of state monopoly to endorsement of foreign-capital participation in the national petroleum industry. Luís Carlos Prestes, PCB leader, said in 1958 that his party considered that injection of such capital into the industry—in the form of assistance to companies or to the government—would accelerate its growth without danger to national sovereignty. A reporter said that "an instant of suspense" followed the statement.[61]

The remaining traditional activists in the "O petróleo é nosso" campaign, however, stood firm in their support. In June 1958 a radical-nationalist slate was elected to the directorate of the Clube Militar. General Justino Alves Bastos, the new president, pledged military backing for the development of Brazil through such means as Petrobrás.[62] In August, radical nationalists won in the National Student Union elections, and the new executive pledged to defend the state oil monopoly.[63]

Abroad, a new crisis erupted in the Middle East in mid-1958, threatening oil supplies to Europe. As had been the case during the 1956 Suez crisis, the possibility loomed that tankers might have to be diverted to the Middle East, prompting *entreguistas* to attack the national oil-exploration effort as inadequate. On the other hand, nationalists warned that "the trusts" would try to use the crisis to "undermine public confidence" in the state oil company; they also said that there was a lesson for Brazil in oil exploitation as allowed by Lebanon and Iraq, and called on the populace to stand firm.[64]

While this crisis simmered unresolved, Argentina surprised the world by opening its oil fields to exploration and exploitation by concession. The United States Secretary of State, John Foster Dulles, came to Brazil just after Argentine President Arturo Frondizi announced the dramatic change in oil policy. Dulles, obviously thinking of Argentina, said at a press conference in Rio de Janeiro that the United States government would neither lend money to Petrobrás nor guaranteee financing arrangements for equipment from United States manufacturers, since private funds were available.[65]

The nationalist press, led by *Última Hora*, charged that Dulles's visit was one more attempt by "the trusts", acting through the State Department, to put down Petrobrás; moreover, the Argentine policy change was the "mask of *entreguismo*".[66] *Última Hora* also reprinted an unfortunate remark from *Hanson's Latin American Letter* of August 2: "the State Department is continuing its policy of backing Brazil against the wall until it has no alternative but to yield as Argentina did." For several months afterwards, *Última Hora* continued to warn of the peril posed by an Argentina in the hands of "the trusts".[67]

The *entreguista* press, headed by Rio's *O Jornal*, greeted Argentina's move with considerable relish, hoping Brazil would be influenced to change its policy.[68] But Petrobrás was apparently a success, and so much rhetoric had been expended in its support that the Argentine policy change had little effect upon Brazilian opinion.

Years of Success, 1954-1960   123

The increase in national oil production, while welcome, presented Petrobrás with a problem. Brazilian refineries (with the exception of Mataripe) were not equipped to handle national crude with its high paraffin content. As a result, most of the national production had to be exported. Early in March 1959 Petrobrás negotiated a contract with the Esso Export Corporation (the international marketing affiliate of Standard Oil of New Jersey) to export 28 million barrels of Bahia crude over three years to Aruba for refining. In exchange, Petrobrás would import 80 million barrels a day of Venezuelan crude from the Creole Petroleum Corporation (also an affiliate of Jersey Standard) over a six-year period.[69]

The FPN became alarmed over this and summoned General Sardenberg for questioning. It charged that the agreement tied Petrobrás to "the trusts" for six years at terms that were not advantageous to the company. Sardenberg retorted that Brazil had no choice but to export her oil, and because of the world overproduction of crude the only bargaining point Petrobrás had was its capacity to import. The company had been fortunate, Sardenberg asserted, to secure such a contract.[70] The FPN did not pursue the matter further.

At this time there appears to have been rather widespread public discussion of Petrobrás, the most enlightened of which had been prompted by an excellent exchange between two articulate spokesmen, representing the two sides. Early in 1957, Roberto de Oliveira Campos wrote an article, "The Three Fallacies of Present-day Brazil", in which he attacked the idea of a state monopoly for the national oil industry and called for liberalization of Law 2004. What he had in mind was not destruction of Petrobrás, but co-operation on its part with foreign oil companies; this would ease the pressure on Brazil's foreign exchange, relieve the expensive burden of exploration, and stimulate Petrobrás by exposure to foreign companies. Campos also lamented the "romantic nationalism" surrounding Petrobrás, which, he said, substituted "units of vanity" for "units of wealth". Finally, he warned that building refineries and ships was a mixed blessing for Brazil. Such activity did indeed save foreign exchange, he admitted, but in the long run; its spur to derivative and secondary industrialization, however, created a new short-run exchange demand, since it contributed to the economic growth of the country. Such subsidiary industries might be long-term savers of foreign exchange, but they were "voracious consumers" of it in the short run.[71]

Later that year, Hélio Beltrão responded in defence of Petrobrás. First in *Última Hora* and then in a pamphlet, "The Six Basic Errors about Petrobrás", Beltrão charged that the interests of Brazil and "the trusts" did not coincide. First, with the current world overproduction of crude, such companies would not feel compelled to make Brazil a producer in the shortest time possible. Further, with Brazil consuming oil to the value of $250 million annually, why should companies who sold it most of the oil be anxious to lose such a market? Brazil was well on its way to self-sufficiency in refining; why should an integrated company be interested only in exploration, the most risky of the sectors? Finally, in Beltrão's estimation, Petrobrás had mobilized "extraordinary" financial resources, which made the sums the international companies could offer "irrelevant". Although there were weak points in both arguments, it is noteworthy that such a high level of debate did go on. The continuing controversy over Petrobrás and state monopoly was thus not always couched in hysterical terms (although it must have appeared so at times).

In mid-1958 a liberal São Paulo newspaper conducted two polls to discover opinion on how Brazil's oil should be developed; the first merely asked readers to choose from a range of options—state monopoly, private Brazilian capital, private foreign capital, Petrobrás and private Brazilian capital, Petrobrás and private foreign capital, private Brazilian and private foreign, or all three. The majority (32 per cent) chose the last, and only 11 per cent favoured leaving development entirely to Petrobrás. The second survey polled professional people who had had (or were having) a university education. Here again, all three sources of capital received preference, with state monopoly far down the list. (It is noteworthy that military officers were polled, and followed the trend.)[73] Although *Última Hora* could (and did) claim with justice that the polls were not representative of all Brazil and that a lot of "free enterprise" types had contributed information, the polls still show that an important segment of the population did not like the way the Brazilian oil industry was being run.[74]

It is striking that not only was there a significant body of opinion opposed to Petrobrás as a means of developing Brazilian oil reserves, but that this segment of the electorate had no influence on Congress. There is no evidence of legislative attempts to have Petrobrás's monopoly curbed. Clearly, Petrobrás was for Congress the only political option available; such had been the case for six years, since the adoption by the UDN of state monopoly had removed the liberal alternative from the debates over the formation of Petrobrás. Clearly, only a massive demonstration of public opinion — on the scale of "O Petróleo é nosso"—in favour of liberalization could have forced Congress to change its position.

In 1959 developmentalist nationalism lost its most articulate spokesman, the Higher Institute of Brazilian Studies (ISEB), as radical nationalists took over its direction. The issue was ostensibly a book by one of ISEB's most distinguished political scientists, Hélio Jaguaribe, that radical nationalists found too pragmatic on foreign capital, particularly on its possible participation in the oil industry. With the resignation of Jaguaribe and his associates from ISEB, radical nationalists expanded their influence (if only by removing a voice of moderation), not just on the oil question, but in the deepening debate over Brazil's economic policy, particularly toward the IMF.[75]

An indication of the deepening polarization of Brazilian politics over development policy came shortly, and Petrobrás was predictably the touchstone. When Roberto Campos suggested that development of national reserves would proceed more quickly if Petrobrás were restricted to the Recôncavo and the rest of the country opened to private exploitation, radical nationalists reacted strongly. The president of the FPN, Bento Gonçalves (of the Republican Party), warned that anyone who "tried to touch" Petrobrás would have a "popular revolution" on his hands. The Chamber of Deputies was in an uproar for several days and the Rio de Janeiro students "declared war" on the president of the BNDE.[76]

In 1959 Walter K. Link, director of Petrobrás's Exploration Department (DEPEX), made a pessimistic evaluation of the petroleum potential of Brazil at the Fifth World Petroleum Congress. *The Oil and Gas Journal* published a condensed version of the paper, but it apparently went almost unnoticed in Brazil.[77] A year later, as he resigned from Petrobrás, Link presented to General Sardenberg a detailed report of six years of exploration in Brazil, in which he made recommendations for the future. Submitted as confidential, the report presented the opinions of the fourteen top geologists (six Brazilian and eight foreign) in DEPEX. Somehow the press got the report, and a tremendous outcry resulted. (This was the first knowledgeable survey of Brazil's petroleum geology that had been widely published.)

Link and his geologists rated each of the twelve major sedimentary basins in Brazil on its probable oil reserves. After expending $300 million in six years of exploration—reportedly the largest single such investment in the world to that time—the only basins about which the department was optimistic were the Recôncavo, Tucano (adjoining the Recôncavo to the north), and Sergipe (the costal strip in the state of the same name). The Middle Amazon, where the discovery at Nova Olinda had been made, was considered marginal; nothing further had been found there. In general, oil prospects in the Paleozoic basins (which comprise 98 per cent of the sedimentary basins of the country) had dwindled as they had been explored. Link concluded by recommending that Petrobrás concentrate its effort in the three most promising basins and that it look for an assured source of petroleum outside the country.[78]

The "confidential" report appeared in several Rio de Janeiro newspapers in mid-November 1960, and Gabriel Passos divulged its contents in the Chamber of Deputies. He charged that Link had "sabatoged" the exploration effort of Petrobrás because secretly he had still been working for Standard Oil of New Jersey. Passos claimed that Link had always ordered "drilling stopped as soon as indications of oil were noted". He even charged that Link had originally been hired by mistake, and that Petrobrás had intended to hire his brother, Theodore, also a prominent geologist.[79] President Sardenberg of Petrobrás hastened to clarify Link's position with the company, and concluded by saying, "The opinion of Petrobrás is that the oil in Brazil is economically utilizable."[80]

At this point, an assessment of Link *vis-à-vis* Petrobrás would perhaps be helpful. He became, and has unfortunately remained, an object of extreme vilification by radical nationalists, and his role with Petrobrás is still clouded by bizarre charges of industrial espionage (ten years later he was still popularly believed to have drilled sideways or not at all!). When Link was hired in 1954, those who negotiated with him made it quite clear that Petrobrás wanted exploration carried out in the same manner as any good commercial oil company. Link approached his task from an economic rather than a strictly geological standpoint: he was looking for fields which could be exploited on a worth-while basis ("commercial" fields, containing over one million barrels of recoverable oil), not just fields.

Moreover, Link was only the head of the fourteen-man directorate of DEPEX, and decisions were made by the group. Apart from those restraints on his "sabotage", the directorate of Petrobrás and the CNP examined the record of Petrobrás every year and certainly would not have approved it had there not been good technical and fiscal reasons for doing so.[81]

All too many Brazilians, however, conditioned by years of reaction to "the trusts", were ready to believe that Link — an American, formerly with Standard Oil, who had challenged long-standing beliefs about Brazil by doubting that it would achieve self-sufficiency in oil production — was a saboteur. If there were any developmentalist nationalists left in Brazil or even in Petrobrás, they seem to have been silent on this issue. Radical nationalism was clearly in the ascendancy, probably because Kubitschek's efforts at fiscal restraint had compromised the appeal of developmentalist nationalism.

At any rate, for the next few years, radical nationalists seized every opportunity to keep Link's name before the public as a "saboteur" and "tool of imperialists". Through speeches in Congress, columns in appropriate newspapers, and the skilful use of other media, radical nationalists held Link up as an example of the danger Brazil constantly faced. Link for his part made only one public statement in his own defence, as he left Brazil. No one else attempted to defend him with much vigour; indeed, it is doubtful if such statements would have been heard in the contemporary wave of reaction.

Just how strong radical nationalism had become may be shown by Sardenberg's testimony in the Chamber of Deputies a few days after the "Link Report" (as it came erroneously to be called — it was, after all, the assessment of all fourteen of the DEPEX geologists) had been leaked. In reply to a question, Sardenberg listed the "pressures" on Petrobrás. The first was "Linkism", which denoted the "defeatist tendency, supported by the shirkers, the weaklings and the cowards". Link, he said, was responsible for another of the "pressures" on the company, since he had tied the entire Brazilian exploration and development program to American methods and equipment; this seriously reduced the flexibility of the company, since it could not negotiate with other countries for equipment and technicians.[82]

While Passos was denouncing Link, another report was made public, which charged that the CNP had permitted the Brazilian private refineries to increase their capacity beyond the prescribed limits and to become indebted to Petrobrás for more than Cr$2 billion. The report had been made by Ernesto Geisel, dismissed from the CNP a month previously in a policy dispute. Article 48 of Law 2004 obliged private refineries to contribute to Petrobrás, and the state company had received slightly more than 3 per cent of its capital from this source by the end of 1959. The three refineries operating on a concession from Petrobrás (Manguinhos, Capuava, Manaus) paid a charge equal to 9 per cent of the CIF value of their crude imports and 50 per cent of their net profits on products refined from imported crude; in addition, all profits earned on refining operations beyond capacity as authorized in mid-1953 had to be turned over to Petrobrás.[83]

The two "plots" became intertwined and FPN deputies, while praising Sardenberg's answers in the Chamber, also campaigned for nationalization of the private refineries and of distribution. The nationalist press joined in calling for "completion of the state monopoly" and denouncing such "internal saboteurs" as Link, the private refineries, and the distributing companies. For example, *O Semanário*, in December 1961, ran the following headline: "Cr$140 Millions was How Much Brazil Paid for Link not to Find Petroleum". Tácito Freitas, writing in the same tabloid in March 1961, charged that Link had had the same role as Victor Oppenheim: to sabotage Brazilian efforts to find oil. Oppenheim had tried to do it in the Recôncavo and had failed; Link had tried the same sinister trick in the Amazon, so Petrobrás should obviously put all its effort into that region.[84] It is significant that not only the radical-nationalist press, but the president of Petrobrás brushed aside the "Link Report" as the work of an agent of "the trusts", unworthy of serious consideration.

In 1960 the New York firm of petroleum consultants, Walter J. Levy, Inc., made a study of Petrobrás's finances.[85] A summary of its findings gives an idea of the principal sources of funds for Petrobrás and of its revenue outlay (see also Table 1). In the first place, nearly 60 per cent of the money made available to Petrobrás up to the end of 1959 was from "external" sources, not from the industry's operations. Of these external funds, fully one-half came from the company's share of the sole tax on locally refined products and of the duty on imported vehicles. Almost 6 per cent came from the compulsory contributions of private refineries. In 1958–59 the company offered preferential shares to the public for the first time, which contributed 0.4 per cent of the total external capital. Secondly, while expansion of operations did provide Petrobrás with an expanding flow of internally generated capital funds, profits were high largely because the CNP set the prices on products from Petrobrás refineries at the same level as on imported products (which were particularly high due to high import duties). National refineries thus had a measure of protection (their products were subject to excise tax, but it was considerably lower than the import duty). This "net profit" accounted for more than two-fifths of Petrobrás earnings in the period. The extent of refinery protection in Brazil was substantially greater than in other countries with similar policies and represented a considerable degree of public support for the company. Moreover, Petrobrás probably was able to get discounts on its crude imports because of world oversupply; this would contribute further to profits. Finally, almost uninterrupted inflation tended to buoy up the company's earnings, but their real value increased much more slowly.

The allocation of funds reflects the fact that Petrobrás had embarked upon the virtual creation of an industry. For example, 35.1 per cent of the additions to gross fixed assets were in the production area, and 40.7 per cent devoted to refinery expansion. Transportation costs accounted for 12.5 per cent of the remainder, in large part for the purchase of four new supertankers in 1958–59. Petrobrás had indicated (without specifying the time period involved) that it had invested 63 per cent of its funds in exploration and development, 19 per cent in refining, and 5.7 per cent for tankers.[86]

General Poppe de Figueiredo (a former president of the CNP) had asserted that every barrel of crude oil represented a saving of $1.[87] The above analysis presents a different conclusion. The Levy study estimated that national refineries received a measure of protection due to high prices on products (approximately 94c per barrel of output), but it was not a "saving". In reality, it represented a subscription of public capital, in the form of government revenues foregone; it also represented a measure of support by the consumer, through high product prices.[88]

Table 1
Petrobrás: Sources of Funds, October 3, 1953 – December 31, 1959

	MILLIONS OF CRUZEIROS	PER CENT OF TOTAL
Internal	17,852	46
Retained earnings	15,074	39
Depreciation and depletion	1,735	4
Provisions for miscellaneous reserves	1,043	3
External	20,630	54
Original capitalization	4,000	14
New capital	16,630	43
Reinvested dividends	2,539	7
Receipts from import duties and for tax on lubricants	12,598	32
Receipts from private refineries	1,390	4
Other	103	
Total Funds	38,482	100

By 1960, *Última Hora* reported that Petrobrás was "receptive" to the idea that the company should exercise a monopoly over petroleum imports. At that time, the private refineries, Petrobrás, and the various distributing companies arranged their own imports. Deputy Barbosa Lima Sobrinho, who had made the proposal for a monopoly, said that the CNP had estimated that a total of $290 million in foreign exchange would have been saved had such a monopoly been operating in 1959 and 1960.[89] In 1960 Brazil signed an agreement with the Soviet Union, for 1960–62, to exchange oil and petroleum machinery for coffee. In the first year of the agreement, Brazil would receive crude oil to the value of $7 million.[90]

In the period covered by this chapter (1954–60), the Brazilian oil industry had indeed grown. Crude reserves increased from 150 million barrels in 1954 to 620 million in 1959 (an indication that Petrobrás had an excellent Exploration Department and that Brazil did indeed have some oil). By 1960 production averaged 85,000 b.p.d., and actually exceeded 100,000 b.p.d. late in December 1960. Plans were being made for Petrobrás refineries in Minas Gerais and Rio Grande do Sul, and for the expansion of Mataripe to 75,000 b.p.d. by 1962.[91]

By 1960, Petrobrás was triumphant. All its sectors—exploration, production, refining, and transportation — had shown dramatic gains in the first six years of its existence. Brazilians had ample basis for optimism about Brazil's future greatness as an industrialized nation, since Petrobrás had apparently shown that the country had abundant supplies of oil, a vital source of energy.

But Petrobrás's horizon was not unclouded. Why, for example, had the company not geared its refineries to process national crude as it was produced? Here was evidence of a lack of executive direction and co-ordination—two sectors working at essentially cross-purposes — which was the result of frequent and politically inspired changes in top management. Then, too, Petrobrás was becoming hypnotized by its own image: it was not trying to discover *if* Brazil had abundant oil reserves, as Brazilians believed, but to *prove* that it *had*. Thus, discoveries were increasingly trumpeted as proof that self-sufficiency was about to be reached. Finally, Petrobrás came to accept and then to preach that it was *the* barrier staving off the rapacious "trusts". Criticism of the company had by 1960 become likened to treason, its sole purpose being to undermine this entity so vital to Brazil's independence. In other words, by 1960 Petrobrás believed that its role was in a sense to confirm the myths about Brazil's oil — that it was abundant and that "the trusts" were eager to exploit it — rather than to investigate them.

6
POLITICAL TURMOIL AND THE STAGNATION OF PETROBRÁS, 1960–1964

In the pattern begun by Janari Nunes, politics became dominant in the administration of Petrobrás in the period 1960–64. Before he took office, Jânio Quadros, the new President of Brazil, was warned from several sides that the company was in a lamentable state. For example, in December 1960 the *Jornal do Brasil*, the least emotional of the Rio de Janeiro newspapers, charged that for three years Petrobrás had been beset with a serious internal conflict, caused by the lack of a definite petroleum policy, structural difficulties in both Petrobrás and the CNP, and administrative incompetence. On several issues (such as the "Link Report") the conflict had become the subject of public debate. Meanwhile, the attitude of the Petrobrás management had degenerated from "flippant optimism" under Nunes, to "bland lassitude" under Sardenberg. Finally, said the newspaper, politicians had begun to interfere in company policy decisions; for example, Kubitschek, campaigning in 1960 on behalf of Marshal Lott (whom Kubitschek's PSD-PTB alliance had reluctantly settled on as his successor),[1] promised Minas Gerais state a refinery of 25,000 b.p.d. capacity. Not to be outdone, São Paulo Governor Jânio Quadros, the UDN-backed reformist candidate, promised Minas a 50,000 b.p.d. refinery.[2]

After the election the "leading technicians" of Petrobrás publicly warned newly elected President Quadros of growing "political meddling" with the company. They said that fewer technically qualified personnel remained in the company's administration after each change in the Presidency of the Republic. But they expressed confidence that "Jânio" would reverse the trend.[3]

The radical nationalists in turn made their demands to Quadros, one of the most important being nationalization of the private refineries. The radical-nationalist deputy, Gabriel Passos, writing early in 1961, said it had become "expedient" to expropriate those enterprises. Petrobrás, he said, had "grown a great deal and will grow more yet"; the private refineries, by law, could not grow, and their proportion of the total refining capacity would dwindle to an insignificant level. Therefore, said Passos, it was better to expropriate now and have done with it.[4] (Conversely, why bother?)

Quadros picked Geonísio Carvalho Barroso as president of Petrobrás. The choice was a departure from tradition in that Barroso was a civilian, a geologist who had been production superintendent for Bahia prior to his appointment.[5] Barroso was chosen in response to pressure from the Bahia oil-workers' unions. Labour in Petrobrás's Bahia installations had not been organized until 1958; to have had such power by 1961 indicates the degree to which Petrobrás had become politicized. According to Glycon de Paiva, the fact that Quadros's choice of Barroso had been "dictated" by labour left Quadros bitter toward Barroso. Quadros's original choice, Josafá Marinho, subsequently became president of the CNP.[6]

In March 1961 the Chamber of Deputies convened an inquiry committee to investigate the "Link Report", which it characterized as an *"entreguista* attack" on Petrobrás. The newspaper *Última Hora* hoped that the committee would show the reforms needed "to fortify Petrobrás against those who wish to destroy it".[7]

In an analysis that Quadros made of the Brazilian exchange situation soon after taking office, he charged that Petrobrás was in "precarious" financial condition. General Idálio Sardenberg, ex-president of the company, denied Quadros's statement before the inquiry committee; he subsequently issued a "Manifesto to the Nation", defending his name and that of Petrobrás. He insisted that the company was "definitively consolidated" and that only "outside forces" could threaten it; that Petrobrás's refining capacity was "sufficient to take care, by itself, of the present consumption of derivatives in the country"; and that "present reserves of crude petroleum are sufficient to maintain for twenty years . . . the production rhythm of one hundred thousand barrels per day, to which I raised it." President Quadros, to discipline Sardenberg and remind the army of its subservience to the Executive, thereupon imprisoned the general.[8]

Sardenberg's claims warrant examination. Walter Link estimated Brazilian reserves at between 600 million and 700 million barrels in 1960; the *World Petroleum Report* sharpened this estimate to 617 million barrels.[9] A production level of 100,000 b.p.d. would exhaust these reserves in seventeen, not twenty, years. Moreover, Sardenberg's assertion that Petrobrás was self-sufficient in refining capacity was also not quite correct: for 1960, Brazil had consumed 250,000 b.p.d; of this total, Petrobrás refineries had processed approximately 80 per cent, and privately owned refineries the balance. True, Mataripe refinery was being expanded from 35,000 to 50,000 b.p.d. capacity, and Duque de Caxias was *supposed* to begin production early in 1961 (as it turned out, it was rather late) with an initial capacity of 60,000 b.p.d., but these developments had not yet come to pass when Sardenberg lashed out at Quadros.

One point that Sardenberg might have mentioned was that Petrobrás refineries were being adapted to process the high-paraffin national crude. Up to that time the production that Mataripe could not handle was exported in exchange for other crudes. This change was due to be realized in 1961.[10] Quite likely Sardenberg remembered the reception the FPN had given an earlier deal to exchange Brazilian crude for that produced by one of "the trusts" and preferred not to raise a potentially damning point when he was trying to build up his former company's prestige.

Radical nationalists outside Congress rallied to the defence of Petrobrás. In June, Lôbo Carneiro declared to the Brazilian Press Institute that "in spite of Link, Petrobrás is in the black!" He also read a statement from the Petroleum Centre which declared that attempts to abrogate the state monopoly of Petrobrás had met "the insurmountable barrier of the resistance of the Brazilian people". Nevertheless, said Lôbo Carneiro, the current campaign by "the trusts" and private refineries indicated that Petrobrás should be "fortified".[11] The nationalists need not have worried, however, because Barroso, in a memorandum to

Quadros, said that Brazilian geologists "diverged" from the conclusions of the "Link Report" and that exploration outside Bahia would continue.[12]

While the proceedings of the inquiry committee were normally secret, with only occasional items released, and the Petrobrás records on the "Link Report" are closed to the public, it is possible to deduce the thinking of company officials on the report. While a few highly placed technicians did indeed publicly charge Link with sabotage in connection with the exploration program, the most common critical reaction was much more moderate. Link, went this charge, was a product of Standard Oil and its methods, used to the luxury of choosing only the best fields and abandoning those which were not exceptional or, more modestly, commercial; Petrobrás could not afford to do so.[13]

The effect of such reasoning was either to reduce the "Link Report" to irrelevance, thus allowing Petrobrás to ignore it, or to assert that it was wrong, which would virtually force Petrobrás to disprove its conclusions. The next few years would show that Petrobrás operated with both goals in mind. The Exploration Department worked largely as if the report had not been made, continuing to scatter its effort throughout Brazil's sedimentary basins; the Public Relations Department seized upon every "show" of oil as proof that Link had been wrong.

All this despite the fact that the report had been a collaborative venture, involving Brazilian as well as foreign geologists, with recommendations made by averaging all opinions. Allowing for the fact that in a few years petroleum technology would advance to a point where the report's conclusions might be less valid, there is no evidence to suggest that Petrobrás was operating at this time (1961) under such a belief. Its reasons were basically political, arising from the fundamental conviction among Brazilians that their subsoil was saturated with oil.

Shortly after taking office and apparently concerned to establish some measure of political support in the northeast, Jânio Quadros ordered the administrative seat of Petrobrás transferred from Rio de Janeiro to Salvador, Bahia. Juracy Magalhães, the first president of Petrobrás and now Governor of Bahia, and a group of "leading citizens" of the state had appealed for the move: Quadros had promised it in his presidential campaign. But the resultant outcry in Rio de Janeiro was so great — one radical-nationalist columnist, for example, charged that the move would bolster "Mister Link's" allegation that the only oil in Brazil was in Bahia—that Quadros "suspended" the plan two months later.[14]

On August 25, 1961, Quadros resigned as President of Brazil, and was succeeded by his Vice-President, João Goulart. The dramatic act shocked Brazil, and Quadros's reasons are still matter for speculation. It may well be that the presidency was too much for him, that he was too easily stymied by an aggressive Congress, and that he lost much of his political following by reversing himself on many questions. Thus isolated, he may have resigned hoping that the army would return him with decree powers rather than have João Goulart, heir to Vargas's following and perhaps another demagogue, as President.[15]

Departing from tradition, Goulart did not replace Barroso as president of Petrobrás. But as Minister of Mines and Energy, he appointed the radical-nationalist deputy Gabriel Passos, who began to prepare the ground for ousting Barroso. First came a vague rumour early in December 1961 that a change in government policy on Petrobrás was imminent. The oil workers of the states of Guanabara (the old Federal District) and of Rio de Janeiro met and declared their opposition to any policy change. Since such a change would mean a new president for Petrobrás, they added that only men "identified with the principles of the state petroleum monopoly" should be considered for the position.[16] At the same time, Lutero Vargas, president of the Rio de Janeiro section of the PTB, organized a rally around the theme "There is no patriotism outside Petrobrás." Representatives of the Brazilian Socialist Party (PSB), the FPN, the UNE and other student organizations, and the staff of *O Semanário* attended.[17]

On December 28 deputy José Jofilly (PSD), acting leader of the FPN, described an interview he had just had with the Minister of Mines and Energy. Jofilly said that Passos intended "to give to Petrobrás a truly nationalistic solution, . . . above any other expediency, personal or regional". Passos considered that "the great enterprise" was at a critical point: it was being attacked by "Linkism" from the outside and by "political appointments and favouritism" from the inside. Its enemies were massing. Jofilly pledged the support of the FPN for Passos, "whose endeavour requires the mobilization of all nationalists".[18]

For another week the Minister of Mines and Energy made frequent dark references to the "problem of Petrobrás", and early in January 1962 implied that the "solution" would come as soon as he had conferred with President Goulart.[19] *Última Hora*, until then a strong supporter of Passos, began to ask what problem Petrobrás was having and why it had come about. On January 3 Paulo Silveira, a columnist for *Última Hora*, charged that Passos was "fighting for the head" of Barroso, and was keeping his motives secret. Silveira said Barroso was one of the best presidents Petrobrás had had, and warned Passos to be careful that in destroying the chief executive he did not destroy the company in the bargain. On January 6, however, Passos removed from office both Barroso and a director of Petrobrás, Heitor Lima Rocha; Passos appointed oil pioneer Irnack Carvalho do Amaral to the company presidency until Barroso's successor could be chosen.[20]

Barroso's dismissal triggered a strike by all Petrobrás personnel in Bahia; the situation was reported "tense", as other unions threatened to strike in sympathy. Irnack Carvalho do Amaral charged that the purpose of the strike was "purely political, having gone beyond the arena of Petrobrás . . . [into] government policy".[21] *Última Hora* criticized Passos for having allowed political considerations to interfere with Petrobrás and charged that he had taken too seriously his role as a fighter for nationalism. Apparently, asserted the newspaper, the government was at the mercy of party feuds and Petrobrás was suffering as a result.[22] Passos did not reply.

While Barroso was the apparent victim of politics, it was not until January 12 that *Última Hora* revealed the reason for Heitor Lima Rocha's dismissal: he and Passos had differed on the oil potential of Brazil. Lima Rocha had submitted a production plan to the Ministry of Mines and Energy in which he had said that production from the Recôncavo could not be raised (it was then at 95,000 b.p.d.), and that the "tendency will be for a decline in the contribution of the Recôncavo fields in the near future." The Ministry had deleted the latter prediction from the report as being "very pessimistic". Lima Rocha had also said that no other sedimentary basin had shown conditions favourable to the accumulation of gas or oil in commercial quantities, even though exploration "had not diminished in intensity." The Ministry had pronounced that statement "so untruthful . . . that it should not form part of an official document".[23] In the light of the "Link Report", the conflict appears to have been between belief and reality about the Brazilian oil potential.

The more extreme nationalists sided with Passos. Adalgisa Nery, for example, a columnist with *Última Hora*, branded Lima Rocha an *entreguista*, and criticized Barroso for refusing to resign over the controversial plan. Barroso, she claimed, had become president as part of a deal by Jânio Quadros with Juracy Magalhães for the votes of Bahia. Barroso, moreover, was to blame for the stagnation of production and of several construction projects.[24]

So many groups forced candidates on Passos that he had difficulty choosing a new president for Petrobrás. Ultimately, on January 15, 1962, he named Francisco Mangabeira; the Bahia strikers expressed their satisfaction and went back to work the same day. Mangabeira at once called for a Petrobrás monopoly over imports of petroleum and derivatives. He said that he would create a department for the workers and that labour would elect a "Social Director" to the company executive.[25]

Mangabeira was a surprise choice; *Última Hora* had not even considered him in its speculations.[26] Apparently Passos had drawn up a list of six candidates after Barroso's dismissal; Mangabeira had not been a serious contender until Passos found that he enjoyed considerable labour support, particularly in Bahia. Mangabeira was a lawyer, a native of Bahia, and a middle-rank executive with the Caixa Econômica Federal of Rio de Janeiro (the approximate equivalent of a trust company in North America). While he was a very able theorist of Christian socialism, he had had no experience whatsoever with petroleum.[27]

In a front-page editorial entitled "Petrobrás to the Wolves" the liberal Rio newspaper *O Globo* denounced the government for its choice. It termed Mangabeira a "truly spiritual figure, so rare in our time that many judge him removed from earthly things", but not the sort of man to tackle the difficult problems connected with Brazilian subsoil exploration. The newspaper also attacked Mangabeira for bringing labour into the administration of Petrobrás, a step hardly calculated to restore public confidence in the company. The government, concluded *O Globo*, had been "cowardly, illogical, irresponsible and two-faced".[28]

Mário Leão Ludolf, vice-president of the Federation of Industries and for ten years industry representative on the CNP, said of the choice, "To be Bahian and a nationalist are absolutely irrelevant for the occupation of the position of President of Petrobrás." He condemned the selection of Mangabeira as "political", asserting that a man of "judgment and experience" should have been chosen.[29]

Adalgisa Nery, on the other hand, called the appointment a "miracle of good judgment". Mangabeira, she wrote, was a serious, humble man, an economist, and a nationalist who was above party politics. In the same issue of *Última Hora*, however, João Pinheiro Neto condemned the government for its "lack of inspiration" in selecting Mangabeira. To move from "management of the real-estate department of a trust company" to the position of supreme administrator of Petrobrás "did not aggrandize him who accepted and much less him who chose."[30]

On May 27, 1962, the *Jornal do Brasil* charged that the dismissal of Barroso had come about as the result of a campaign conceived and directed by Gabriel Passos and co-ordinated by Air Force Colonel Anderson Mascarenhas, a radical nationalist. In the crisis that followed, the government had moved to the Catholic Left to pick a compromise candidate. Mangabeira was apparently the only candidate to whom no political faction could object, and so became president of Petrobrás by elimination. The rallies held early in December, pledging firm support for Petrobrás, had probably also been part of the campaign engineered by Passos to guarantee popular support for internal reforms made in the name of nationalism. Mascarenhas, in payment for services rendered, the newspaper charged, became commercial director of Petrobrás.[31]

Charges of corruption in connection with Petrobrás were becoming more and more common in the early 1960s. Many centred on the Duque de Caxias refinery, whose inauguration had been delayed for a full year by strikes and alleged sabotage. For example, production costs at Duque de Caxias were rumoured to be two to three times higher than normal. The refinery, it was reported, operated with forty-nine employees per thousand barrels of capacity, as opposed to the privately owned União refinery at Capuava, São Paulo, which operated with eighteen per thousand barrels.[32]

Such charges were made, not just by Petrobrás's opponents, but by radical nationalists who were concerned that the company might weaken itself for the continuing battle with "the trusts". The radical-nationalist columnist Adalgisa Nery frequently wondered where the Clube Militar was when Petrobrás seemed to need it most.

A clue emerged in a statement from the Minister of War concerning the 1962 elections in the club. General Segadas Viana, a former president of the club, decided to quell the growing factionalism in the organization by stopping all discussion of national problems in speeches and in the club's *Revista*. He emphasized that the institution could not "in any way want to impose solutions whose responsibility [did] not fall to [it], but to the government." The directorate, he concluded, should concern itself with the principal activities of the club, "which are of a social, cultural and recreational nature."[33]

Not that much had been heard from the reputedly radical-nationalist directorate, led by General Justino Alves Bastos and elected in 1958. By that time the club's organ, the *Revista do Clube Militar*, was being published only sporadically (indeed, it had not appeared for some months). The military as a bloc and the Clube Militar in particular appear to have deliberately "maintained a low profile" during Kubitschek's term. The reasons are not entirely clear, but may relate to the 1955 plot — that Lott foiled — on the part of anti-Vargas elements within the military, and linked to a civilian bloc led by Carlos Lacerda, to prevent Kubitschek's inauguration. The military appears to have closed ranks as a result, and tried to stay out of potential trouble in order to rebuild its compromised image as a moderating force.[34] It would appear, then, that Segadas Viana was trying to maintain this attitude and forestall the real possibility that the club—

and the military—might be rent by open strife between nationalists and *entreguistas*.

By early 1962 a new emphasis was apparent in the publicity about Petrobrás: for some time company officials and supporters had stressed the stimulus Petrobrás was giving the national economy through purchases within the country, and had passed over the stagnating production picture.

Such stories pointed to opportunities created by the growth of Petrobrás. For example, in expanding the Mataripe refinery of Bahia, 25 per cent of the material used had been of national manufacture. By contrast, 60 per cent of the material used in the construction of the Duque de Caxias refinery had been made in Brazil.[35] In 1961 Petrobrás had for the first time awarded a contract to a Brazilian company, in that instance to build three new processing units for the Cubatão refinery. The "greater part" of the necessary equipment would be bought in Brazil. In 1962 Petrobrás ordered six tankers from Brazilian yards; the first, the *Jacuipe*, of 10,500 dwt, was launched in June 1964. Petrobrás was purchasing drilling rigs made in Brazil by 1964, and by that time was buying close to 85 per cent of its equipment within the country.[36]

Occasionally, Petrobrás mentioned the benefits its activities brought to Brazil in other ways. For example, the company said that exploration in the area had changed Nova Olinda from "a couple of shacks on the riverbank" into an "authentic city" with paved roads, electricity, water and sewers, a medical post, and a primary school, "all constructed by Petrobrás". Some of the wells Petrobrás had drilled brought in water for a particular region, rather than oil. Roads constructed as part of the exploration effort had helped the nation.[37]

These "benefits" warrant examination. By Article 27 of Law 2004, Petrobrás was obliged to pay an indemnity of 5 per cent of the value of the product extracted (reduced to 4 per cent in 1957) to the states and territories in which the company was producing oil. Of that "royalty", 20 per cent had to be turned over to the *municípios* (an approximate equivalent of the Anglo-Saxon county) in which such production was taking place, divided proportionately. In practice the states got the "royalty" in cash, and the *municípios* received their share of it in public works and services contributed by Petrobrás. In a sense, then, Petrobrás's publicity was making a virtue of necessity.

Despite such attempts to justify its performance (and perhaps to distract attention from stagnating production), not all critics were convinced. Mention has already been made of the charges of corruption that were increasingly bothering supporters and opponents of the company, but the solution proposed by more and more radical nationalists is of interest here. Led by the company's president, Francisco Mangabeira, such individuals were pressing for labour participation in the direction of the company, coupled with the institution of a broad program of social benefits.[38]

Another solution to the problem came from Minister of Mines Gabriel Passos, who put the blame for all Petrobrás's troubles on "Linkism". It was all part of a conspiracy by "private interests" to thwart his efforts on behalf of Brazil (a charge similar to the one Vargas had made in 1954 and Kubitschek had reiterated in 1957). As a solution, Passos planned to divide Brazil into exploration zones; the work in each zone would be done by technical groups composed of foreigners ("each one of a different nationality") and a fixed percentage of Brazilians.[39]

But Petrobrás was able to redeem itself somewhat with the announcement that a wildcat well had yielded oil at 5,740 feet in Bahia's Tucano basin, which the company estimated was larger than the neighbouring Recôncavo. Such nationalists as Adalgisa Nery and Paulo Silveira hailed the discovery as the definitive destruction of the "Linkistas" and *entreguistas*. Francisco Mangabeira said Link had not explored Tucano until 1960, when he had drilled three dry wells. Despite such "pessimistic" results, Mangabeira declared, Link had concluded that Tucano was one of the most promising regions in Brazil (backhanded proof of Link's incapacity?).[40]

A month later Petrobrás claimed to have discovered oil at Tabuleiro dos Martins, Alagoas, and the nationalist press heralded yet another victory for the company over the pessimism of the "Link Report". Columnist Joel Silveira, of *Última Hora*, forecast self-sufficiency in crude-oil production in the near future; the field, however, turned out not to be commercial.[41]

Early in 1962 a group of Brazilian industrialists submitted a plan for the development of national oil shales to President Goulart, who referred it to Gabriel Passos. The plan called for the construction of a pilot plant at Irati, Paraná, for distillation of the shales; financing would be through a loan of $7 million from the United States. In its story the *New York Times* estimated that a pilot plant with a capacity of 12,000 b.p.d. would cost $88 million. A plant having ten times that capacity would cost $600 million.[42]

A few days later Passos announced that the Brazilian government was interested in the development of oil shales, and would try to negotiate financing with Russia and the United States. (Since the election of John F. Kennedy, the United States had become willing to give loans for government, not just private, enterprise.) Passos said Petrobrás was constructing a prototype distillery at São Mateus, Paraná, and technicians of the state company would survey the Irati shales.[43] The Minister of Mines and Energy clearly could not allow private capital, probably linked to "the trusts", to gain control of part of the national oil industry.

Through the Alliance for Progress, the United States offered to lend Brazil $7 million for oil-shale development. The loan was for seven years, in the hope that by that time the operation would be producing 250,000 b.p.d. If production were not at least 75 per cent of the desired figure by 1970, repayment of the loan would be deferred for forty years. But it was stipulated that a United States technician should be in charge of the operation, "to guarantee the success desired".[44]

The entire proposal became a political issue in Brazil. Adalgisa Nery condemned the loan as "just another attack on Petrobrás". Since the operation was to be supervised by an American, she charged, he would make sure that the plant did not reach 75 per cent of the desired production level and that the United States would take it over in forty years. The technician would have the same mission as Link had had, she said: to hold back Petrobrás. The Soviet Union had also made an offer to finance development of the shales, which only served to make the controversy more heated.[45]

No sooner had this outcry abated when a new conflict appeared, over the purchase by Petrobrás of liquid petroleum gas

(LPG) from Argentina. In December 1961 Petrobrás had broken a contract for LPG with "Mundogás", an Argentine subsidiary of Standard Oil of New Jersey, to negotiate directly with YPF. Mundogás had offered LPG at $73 per ton, while YPF offered it at $68, payable in local currency. By the YPF contract, half the transportation would be through the Brazilian tanker fleet, and Argentina agreed to purchase Brazilian manufactured goods to the total value of the LPG sold to Brazil. Argentina, however, wanted to have an intermediary company, FAROS, Sociedad Anónima, representing YPF. Petrobrás had always dealt directly with the supplier, and negotiations lagged until Mangabeira accepted the contract with FAROS, against both the opinion of his advisers and company policy.[46]

Another foreign contract stirred controversy at the time, but with a Brazilian company. In February, two Brazilian financiers, Celso da Rocha Miranda and Mário Wallace Simonsen, formed Petronal-Petróleo S.A. to explore in the Persian Gulf. The company, acting as an intermediary, also proposed supplying Petrobrás with 120 million barrels of Persian Gulf crude over five years. Payment would be partly in dollars, partly in cruzeiros, with the dollar proportion diminishing over the length of the contract. In a description published in May, the company claimed that its members had been active in companies which had explored in Paraguay and Bolivia.

The published description was in answer to bitter denunciations of the contract not only by radical nationalists but by various financial officers of Petrobrás. Again, the proposed deal was with an intermediary, and Petrobrás's policy was to deal exclusively with producers. Radical nationalists claimed that the company was linked to the Chase Manhattan Bank, the Rockefeller "group", British Petroleum, and other "trusts". Mangabeira's action over the proposed contract appears to have been the most provocative point: he had tried to rush the contract through Petrobrás's administrative council without first submitting it to technical scrutiny, and the move had only just been caught by one of the company's councillors. It appears that the contract was never approved.[47] At the same time, Mangabeira was called before a Congressional Inquiry Committee to discuss the operations of Petrobrás. One question concerned the delay in supplying fuel to the Brazilian Navy. Mangabeira blamed his commercial office, which he said had created so many obstacles that the navy had finally appealed to Mangabeira himself.

Adalgisa Nery had seen a campaign being mounted against Mangabeira because he had made it possible for Petrobrás to supply the navy. Up to that time, the "vast profits" of the supply had been pocketed by a "powerful and world-renowned petroleum company". Petrobrás supplied fuel to the U.S.S. *Constellation* when she stopped in Rio de Janeiro, an event which caused some expressions of pride in the nationalistic press. The decision to supply the fuel requirements of the navy was, according to Mangabeira, the first step in a plan to remove importation and distribution from the hands of the "international petroleum cartels". During 1962 he pursued this policy methodically, and Petrobrás began to supply a growing list of public entities with fuel.[48]

The internal divisions of Petrobrás, caused by Mangabeira's mismanagement, came to a head on May 24, 1962, when the directors of all the technical sections (followed two days later by the remaining engineers and technicians) demanded the president's dismissal. Mangabeira appealed for support to the company's labour syndicates; they agreed, but demanded conditions: the absorption of all private refineries into Petrobrás, a Petrobrás monopoly over the importation and distribution of petroleum and derivatives, and the abolition of the CNP. Mangabeira agreed to press for the demands and, with labour backing, dismissed all the directors of technical departments.[49]

The dismissed and striking technicians then published a revealing "Manifesto to the Nation" on May 26. They demanded Mangabeira's dismissal, a purge of all Petrobrás personnel who did not support state monopoly, completion of the state monopoly, and the participation of labour in the management of the company. Only with such a united direction, the document concluded, could Petrobrás survive the attacks of international groups and their Brazilian allies.[50]

What appears to have happened is that Mangabeira had become intolerable to the Petrobrás management, and the company's labour unions had become so influential that both the president and his managers immediately turned to them for support. The only difference between the conditions the syndicates presented Mangabeira and the subsequent manifesto that the technicians published was that the latter demanded Mangabeira's dismissal. The manifesto wooed labour by demanding its inclusion in company management; the syndicates, if Mangabeira survived, obviously had or would get such a commitment from him.

Analysing the crisis, the *Jornal do Brasil* traced its origins to the administration of General Idálio Sardenberg, who had encouraged agitation by shortening working hours and yielding to pressure from both regional groups and the "Left". Mangabeira had continued such practices, and had granted "substantial" salary increases for all employees. "Leftist elements" had begun to take over the company under Mangabeira, according to what the newspaper called "the best and most experienced technicians". The newspaper charged that Mangabeira had assumed leadership of the forces determined to blame Link for the lack of oil in Brazil. Whether the campaign discovered oil or not, "it could furnish votes in the elections, in Bahia and in other states." Mangabeira was ten years behind the times, asserted the *Jornal do Brasil*; he was campaigning for "O petróleo é nosso" and did not perceive that the state monopoly was law. In trying to bring men of "leftist" ideologies into the company, he had made some disastrous appointments, and was trying to hide his incompetence by carrying his case to the public, accusing the striking technicians of "idolizing the cartels" (which, according to their "manifesto", could not have been further from the truth).

Mangabeira, pursuing a tough policy, wired the Petrobrás units in Bahia that resignations offered "for any motive" up to the level of superintendent and unit chief would be accepted. At the same time, the *Jornal do Brasil* reported, he approached General Segadas Viana, Minister of War, asking the army to supply technicians to fill the many vacancies (rumour put the number at between 175 and 200), but the Minister refused.

On May 30 the same newspaper reported that no less a figure than General Arthur Levy, superintendent of Duque de Caxias refinery, had resigned. The Superintendent of Cubatão refinery, Oto Martins de Lima, also was reported to have resigned.

The workers were apparently calling for the dismissal of several other men high in the administration of Petrobrás. From the reports, it appears that there were dismissals as well as resignations, from which one may infer that labour had cast its lot with Mangabeira, thus encouraging him to a display of bravado (he dismissed many of the men who had signed the memorandum calling for his own dismissal).

On June 1 Mangabeira was in Brasília actively seeking the support of radical-nationalist deputies. He maintained that opposition to him was "of little significance" and was crumbling; many of the 175 men who had resigned had changed their minds and wired him their support, he said. The administrative council of the company thereupon banned news releases from Petrobrás. Then, on June 20 Petrobrás revealed that each replacement for a man who had resigned would be hired to fill a variety of posts rather than for one specific position (with the exception of the most important vacancies, which had already been filled).

The death in June 1962 of Minister of Mines and Energy Gabriel Passos further confused the situation. The *Jornal do Brasil* said that Mangabeira would remain president of Petrobrás at least until a new cabinet was formed. Meanwhile, it added, Petrobrás was virtually paralysed, owing to bitterness over the changes in personnel. (Apparently, the technicians of Duque de Caxias refinery had refused to accept their new superintendent, the staff of the Consultoria Econômica had refused to co-operate with Mangabeira, and the company was in turmoil over the crisis.)

On a visit to the Recôncavo oilfields, Mangabeira intimated that he was interested in becoming a candidate for the federal Chamber of Deputies, as soon as he had been released from Petrobrás. Three days later, while still in office, he announced he would contest the 1962 election as a PSB candidate. He also declared he would not press for expropriation of the private refineries; Guanabara syndicate leaders accordingly announced they were "not disposed to fight any longer" to keep Mangabeira in office. Apparently, despite his backtracking on the private refineries, he did not find much support for his projected congressional candidacy. Three days later, the *Jornal do Brasil* reported that Mangabeira was working to stay on as president of Petrobrás.[51]

Mangabeira had obviously alienated labour by backing down on expropriation of the private refineries (probably because the Goulart administration had intimated that it could not afford to expropriate, prompting Mangabeira to reverse himself to preserve his position), and he would not have been able to count on labour support at the polls. On the other hand, he had eliminated considerable personal opposition from Petrobrás and could have stayed on because it would have been very difficult to find a successor for him at such a troubled stage in the company's history. Like the government he served, Mangabeira was floundering about in search of personal survival.

These crises had their effect upon output. Early in August, Mangabeira was asked in the Chamber of Deputies why production had declined from 95,000 b.p.d. in 1961 to 90,000 in the first six months of 1962. He blamed "superabundant rains" in the Recôncavo early in the year, which had chilled the tanks, causing the oil to condense in them, and which had also impeded communication with offshore wells.[52] The Petrobrás Annual Report for 1962 blamed the January strike, and said that the company had not been able to reverse the declining trend until October, when some long-awaited equipment finally arrived from the United States (by implication, Americans were again at least partially to blame for the difficulties Petrobrás was encountering).[53]

The *Jornal do Brasil*, however, blamed the "administrative crisis" in Petrobrás. For example, it said, not one new drilling rig had entered service in the first half of 1962, and of the sixty-one such units held by the company, thirty had been under repair and one declared unserviceable. Accordingly, fewer wells had been drilled and, since not all defunct wells could be replaced, production had declined. Furthermore, the *Jornal do Brasil* charged, the general production level had dropped because of poor oil-extraction technique.

The newspaper quoted statistics to support its charges. For example, the monthly drilling average had been 18.9 wells and 24 kilometres in 1959, 20.7 wells and 28.5 kilometres in 1961, but had fallen to 16.7 wells and 19.3 kilometres in the first half of 1962. Similarly, production had risen from a daily average of 64,630 barrels in 1959 to 95,363 in 1961, but had declined to 87,261 barrels in the first half of 1962.[54]

Although Mangabeira did not run after all in the 1962 elections, he played a part in them. As president of Petrobrás, he threw his support (and by inference the support of the oil company) behind certain candidates. Late in September, for example, he publicly recommended PSB candidate Max Costa Santos for the Chamber of Deputies. He praised Santos as a "nationalist of long standing" and a "defender of a monopoly for Petrobrás on the refining and importing of petroleum", who merited "the esteem and the confidence of all who fight for the great state petroleum company, our Petrobrás". The election of Santos, said Mangabeira, would reinforce the nationalist bloc in Congress, "whose banner is the economic and social emancipation of the Brazilian people".[55]

He later denied using Petrobrás funds for political or electoral purposes, but a Congressional Inquiry Committee nevertheless recommended that Petrobrás be restrained from using its revenues for "ideological" ends. (Eugênio Gudin charged that Petrobrás had spent Cr$150 million on the 1960 elections, promoting "national communism" candidates.) The committee also called for "strengthening" of the state monopoly as "an essential instrument in the struggle for economic emancipation".[56]

By January 1963 the growing polarization of Brazilian politics had begun to involve Petrobrás even more deeply, and *Última Hora* published a series of articles calling upon nationalists to "reassert their faith" in the company to thwart *entreguista* attacks. Leading nationalists contributed material, and the series ran for the last three weeks of January. The series included articles by such men as General Horta Barbosa ("Without the State Petroleum Monopoly, We Will Not Arrive at Economic Emancipation!"), General Teixeira Lott ("Defence of the State Petroleum Monopoly Is a Patriotic Duty of All Brazilians"), General Osvino Alves, commander of the First Army ("The Armed Forces Defend the State Petroleum Monopoly"), and other stories in the same vein.[57]

In the same month, the CNP announced that Petrobrás would supply all government organs and "semi-public companies"; moreover, the CNP was studying the national consumption of derivatives, in preparation for Petrobrás taking over distribution.[58] Petrobrás had already entered the distribution sector, having been given permission to do so by the CNP in August 1961.[59] The company inaugurated its first service station in Brasília on June 27, 1960 — presumably on a test basis and as a patriotic gesture to the just-inaugurated capital city, since the CNP had not then authorized the company to expand into distribution. The second retail outlet opened in 1962 near São Paulo.

At the same time, owing to a growing deficit in foreign exchange, the government removed the special exchange rates for imports of petroleum. The practice of setting exchange rates to make cheaper the importing of essential products (oil and wheat were consistently accorded such treatment, with other commodities added when the need arose) had begun under Kubitschek. They were, in effect, a government subsidy for the industry, and controversial. Liberals abhorred them — Jânio Quadros had removed them upon entering office, only to restore them a few months later, as he embraced nationalism — and nationalists of both types defended them as a necessary initial means of support for such fledgling and essential enterprises as Petrobrás. They were expensive for the government, since they robbed it of precious foreign exchange.[60]

In mid-1963 Francisco Mangabeira resigned from Petrobrás. He said he was leaving the company "doubly satisfied": he had served in the best way he could for the economic emancipation of Brazil and had "brought Petrobrás down from its ivory tower". He said Goulart had offered him several other positions, including the presidency of the CNP, but he had refused them all because he needed a holiday.[61] In all likelihood, Goulart had waited to replace Mangabeira until the controversy over his mismanagement had died down and until he could make the substitution without appearing to have yielded to pressure.

Mangabeira's successor, General Albino Silva, Chief of Goulart's Military Household, was an unknown quantity. Petrobrás's unions had threatened to strike if a nationalist were not appointed, but decided to postpone action until they had been able to form an opinion of Albino Silva. Mário Lima, head of the Refinery Workers' Syndicate and a federal deputy, warned the new president that he faced a general strike if he were "to throw his weight around". Lima expressed regret that Mangabeira had left, but said he was pleased that Jairo José Farias had become a director of Petrobrás. Hugo Régis dos Reis was also made a director; both appointments had been "indicated by the workers of the company". Farias had previously been superintendent of Mataripe refinery (since June 1962). The "producing classes" (liberal-minded businessmen) of Bahia protested to Goulart about the appointment of Farias, charging that he had defended "anti-Christian, anti-democratic ideas" while at Mataripe and that he would pose a threat to national security. The force of the accusation was weakened somewhat since the petitioners said they were upset over the appointments because Bahia had been excluded from the directorate of Petrobrás.[62]

Albino Silva's appointment must be seen in the wider context of Brazilian politics at the time. By mid-1963 the country was, to put it bluntly, in a state of crisis, brought about by Goulart's weakness and indecision as president. Goulart seems never to have been taken seriously, and was held to be admirably suited for no more than the vice-presidency: he could rally enough of Vargas's PTB following to help elect a president (he had done so for Kubitschek and Quadros), but he had no leadership capabilities, at least not of the quality required for the increasingly polarized Brazilian political scene. Radical groups on both the right and left were attracting growing numbers of adherents, disenchanted with Goulart's floundering. By mid-1963 these groups were openly discussing *coups*; moderates were wondering if Brazil could survive until 1965 and the end of Goulart's mandate. Such indecision on the part of Goulart served only to strengthen the hands of the extremists. In such a context, Goulart tried to appease the group that was most crucial to his survival, the army. No *coup* could succeed without their active support, and Goulart shuffled his cabinet in June to include more representatives from the armed forces. Albino Silva's appointment was of this sort.[63]

In his June 1963 inaugural address, Albino Silva promised to resist "the cupidity of the international trusts", and to pursue the policy of his predecessor. (Considering Mangabeira's equivocation over such questions as expropriation of the private refineries, he left himself considerable scope.) After Albino Silva's address, the FPN announced its satisfaction with the choice.

João Pinheiro Neto of *Última Hora* also was satisfied, but called on Albino Silva to revamp the company's organization and inject austerity and renewed efficiency into it; he said there was no more need of "grandiloquent manifestations". The *Jornal do Brasil* was unhappy with Albino Silva's speech, since he had made no mention of exploration or production and had laid such emphasis on the "personal" goals of his predecessor. The newspaper charged editorially that the company had become a "sink of corruption", and that a thorough reorganization was needed. One indication of Albino Silva's sympathies was that he had denounced the Roboré Accord on nationalistic grounds in a speech to the Clube Militar in 1959. The new president differed from his predecessor in one significant aspect: Albino Silva's name virtually did not appear in the press while he held office in Petrobrás. The company stayed in the news because it pursued Mangabeira's program, and Albino Silva apparently exercised little influence over company policy during his term of office.[64]

In the following month, Carlos Lacerda, Governor of Guanabara state and an outspoken critic of Petrobrás, toured Rio Grande do Sul and was met by violent labour demonstrations in Pôrto Alegre. He charged that Petrobrás employees had been the "focus of the disorder" against him, which proved that the company had been "handed over to the Communists". Albino Silva, Lacerda concluded, was acting as a "façade" for Communist infiltration.[65]

The employees concerned were engaged in the construction of a refinery near Pôrto Alegre and worked for Petrobrás itself, not a contracting firm. The incident demonstrates how politicized Petrobrás labour had become. Since Rio Grande do Sul was the home state of Leonel Brizola, Goulart's brother-in-law and perhaps the most demagogic of contemporary radical nationalists, and since Lacerda was the personification of *entreguismo* to radical nationalists, it was probably inevitable that some nationalistic group would demonstrate against him.

Deputy Peracchi Barcellos (PSD), of Rio Grande do Sul, bitterly denounced Petrobrás for the incident. He said that like all true patriots, he had "defended" the company; never had he imagined that it could be transformed into "an agent of disorder and riot in this country". Barcellos charged that employees of

Petrobrás had been mobilized to promote unrest upon the arrival of Lacerda; the chief of the state police had identified Petrobrás workers directing the demonstration.[66]

Next, Deputy Ortiz Borges (PTB), of the same state, replied to these charges, quoting an article that the Employees' Association of Petrobrás (of Rio Grande do Sul) had inserted in the Pôrto Alegre newspaper *Correio do Povo*. The decision to participate in the demonstration against Lacerda, claimed the labour group, had been taken "unanimously, in an extraordinary general assembly", and had been "dictated" by the employees' firm position of "consistent struggle in favour of state monopoly and against imperialism and its agents". The demonstration had been orderly and the chief engineer had "democratically" respected the unanimous decision of the workers, who had been exercising their constitutional right to freedom of expression. Finally, the article called, "in the name of the students, workers and the people in general", for the repudiation of the agents of international interests, for the struggle for basic reforms, and for an integrated petroleum monopoly (starting with the expropriation of the private refineries).[67]

By mid-1963 Petrobrás was in financial difficulties, and its recent drive toward completion of the state monopoly was at least partly to blame. Petrobrás had taken over the distribution of derivatives to public organizations in January, but the company had not considered that government departments might be slow in paying their accounts. By August federal agencies owed Petrobrás Cr$16 billion (almost $27 million), and the total was steadily rising. The private distributing companies had had the same problem with the Brazilian bureaucracy, but had always paid Petrobrás in thirty days. By October, however, the situation had apparently improved somewhat, since sixteen government entities owed a total of only Cr$1.5 billion. They included the Ministry of the Navy, NOVACAP (the agency in charge of construction in Brasília), and Loide Brasileiro (the national shipping line). At the same time, Petrobrás was several months behind in "royalty" payments to Recôncavo *municípios*, and owed about $100 million to foreign suppliers of petroleum.[68]

Despite these difficulties, pressure for expropriation of the private refineries intensified. Albino Silva declared that such a move would be "just", but stressed that the decision had to be made by the President of the Republic, Petrobrás being a "mere executor of the state monopoly". Workers at the Capuava refinery of Refinaria e Exploração de Petróleo União were apparently ready to strike to force nationalization of their plant and, if they did, the Petrobrás labour force might walk out in sympathy.[69] By now the government, too, was interested in expropriation, as a sop to the extreme nationalists (led by Leonel Brizola), who were attacking the Goulart administration over quite another matter.

Radical nationalists had been angered by what they saw as a "sellout" by Goulart and his Finance Minister, San Tiago Dantas, over the purchase price of the Brazilian holdings of the American and Foreign Power Company (AMFORP). Goulart had visited the United States in April 1962, and had negotiated a purchase price of $135 million; a year later AMFORP and the Brazilian government signed an agreement at this price. The nationalists charged that the settlement was too generous to the company; furthermore, the announcement came in April 1963, in the midst of an impassioned debate over a bill on civil-service and military salaries. The radical nationalists bitterly opposed the settlement with AMFORP — Skidmore writes that the opposition "frightened" Goulart — and the government turned to expropriation of the private refineries as a means of soothing the radical nationalists.

In an editorial entitled "National Distortions", the *Jornal do Brasil* said the campaign for the nationalization of the private refineries illustrated what was happening in the country. The administration of Petrobrás did not lead, but was at the whim of pressures from below; the administration of the country was in the same position, charged the newspaper.[70]

Then, however, Brizola reversed his policy and denounced expropriation. He claimed that foreign distributing companies were behind the nationalization campaign, because they would rather deal with Petrobrás in its weakened condition than with the — by implication, stronger — private refiners. The state company had "empty coffers and would not refuse the hand extended by the companies who can offer credits and long time periods" for payment. Privately, Brizola was worried that the nationalization campaign might backfire, creating a "climate of insecurity" which would permit a "strongman" advocating private enterprise to rise to power by means of an "unconstitutional adventure". He apparently considered even João Goulart a possible strongman.[71] *Última Hora* also denounced nationalization of the private refineries as "inappropriate" at the time. The government, said the daily, was striving for "basic reforms", and expropriation at that moment would prejudice overall development and hinder the reform effort.[72]

Just by the way, if nationalization were being considered solely for economic reasons, it was unnecessary. All Petrobrás had to do was purchase a majority of the shares in each private company to achieve control of the refineries. But economics appeared to have little to do with the nationalization of Brazil's private refineries at this time.

On the other side, the Engineers' Association of Rio de Janeiro and Guanabara called on September 9 for the expropriation of Capuava refinery, pointing out that not only would the São Paulo area be faced with a shortage in refinery capacity within three or four years, but because of space limitations Petrobrás's Landulpho Alves refinery (at Cubatão) could not expand. It would be cheaper to nationalize Capuava (which could grow) than to build another Petrobrás installation in the area.[73]

In November *Última Hora*, clearly swinging behind Brizola, attacked the growing labour campaign for expropriation as "unpatriotic" and "of unequivocally Communist inspiration". It charged that Communists had "taken over the key posts of Petrobrás" and had prejudiced basic national reforms. They were trying, said the newspaper, not only to expropriate the private refineries, but Petrobrás itself and its workers, as well as students and intellectuals.

Then Waldir Pires, the Solicitor General of Brazil, said that a measure for expropriating the nation's private refineries had been prepared for President Goulart's action. Pires said he believed the President would act when a "favourable political situation" appeared. The employees of Capuava were then on strike in support of expropriation.[74]

In the middle of the month ten thousand employees of Petrobrás went on strike for twenty-four hours to force the nationalization of Capuava refinery. They suspended their walkout only after Minister of Labour Amauri Silva promised that President Goulart would expropriate the refinery at an opportune moment. Albino Silva, after a conference with Goulart, said that the government would not tolerate a general strike of petroleum workers. The First Army had taken up positions around refineries in the Rio area, and the government clearly intended to remain firm against labour pressure. The Minister of Labour also reported that all demands made by the striking employees of Capuava had been found acceptable, and the strike there would soon be over. Curiously, no mention was made of expropriation in the *Última Hora* account of the Capuava workers' demands; only salary and benefits were apparently at issue.[75] Either this was a reflection of the newspaper's new policy against expropriation, or the workers had changed their minds.

Four days later a major confrontation over nationalization of private refineries was postponed when the General Labour Command (CGT) called off a threatened general strike to force the immediate expropriation of Capuava refinery. The CGT said that although the Petrobrás technicians were in favour of a strike, "public opinion and the majority of workers are not yet prepared for a struggle in terms of a general strike."[76] There matters rested for the time being.

But on December 24, 1963, the government took another step toward completion of the state monopoly on petroleum by granting Petrobrás a monopoly over imports of crude oil, which would give the company more bargaining-power in negotiations on the world market, since it imported oil for its own refineries as well as those owned by private Brazilian capital. Three days later the *Jornal do Brasil* published the following breakdown of the countries of origin for crude supplies of the various Brazilian refineries:

> Capuava (30,000 b.p.d. capacity) — 20,000 b.p.d. from Kuwait (Gulf Oil) and 10,000 from the U.S.S.R.
> Manguinhos (10,000) — all from Venezuela (Standard Oil)
> Sabbá (Manaus) (5,000) — 4,000 from Venezuela (Shell),
>   1,000 from Peru (Sinclair)
> Ypiranga (9,500) — all from Venezuela (Gulf and Standard Oil)

Cubatão (110,000) — 40,000 from Venezuela (Sun Oil),
10,000 from U.S.S.R., 45,000 from Bahia,
15,000 from Venezuela (Shell)
Duque de Caxias (100,000) — 10,000 from France (state oil company),
15,000 from Bahia, 75,000 from Saudi
Arabia (Texaco) and Venezuela (Shell)
Mataripe (42,000) — all from Bahia.

Petrobrás signed its first contract under the monopoly legislation on March 7, 1964, and the then president, Army Marshal Osvino Alves, claimed that the company would save $2.5 million per week as a result of the monopoly (the figure has proven impossible to verify). The *Petroleum Intelligence Weekly* later asserted that Brazil had become a "barometer" for world crude prices, because of the sheer size of the market Petrobrás controlled.[77]

Meanwhile, a new dimension had been added to Brazil's oil controversy. Late in August 1963 Petrobrás director Hugo Régis dos Reis revealed that two Soviet petroleum technicians were in the country and that they had concluded that "Brazil possesses petroleum in great quantity and in commercial terms." They had advised intensification of drilling. The *Jornal do Brasil* suspiciously asked why, since Brazil maintained diplomatic relations with the Soviet Union, nothing had been divulged about the coming of the technicians. What should have been a normal event, the newspaper said, had been "shrouded in unjustifiable mystery". Moreover, the full report had not been divulged, nor had there been any comment on the apparent discrepancy between the new assessment and the "Link Report".[78] The Russian interest was confirmed in October, when Albino Silva reported that the two Soviet geologists had indeed arrived some six months earlier; they had stayed a short time and had produced a report based entirely on Brazilian data.

The visitors had apparently asserted that Brazil could quadruple its oil production within five years, thereby achieving self-sufficiency in crude-oil output. *Última Hora* predictably welcomed the "Russian Report", asserting that it "destroyed the pessimism of the Link Report" and showed the truth about Brazil's oil potential. But, curiously, the newspaper did not publish anything further about the visit.[79]

Apparently, the only newspaper to carry a lengthy presentation of the "Russian Report" was the radical-nationalist weekly *O Semanário*. In this article, General Tácito Freitas described most of the Russians' discoveries and conclusions about Brazil. Petrobrás, according to Tácito Freitas, had hired Bakirov and Tagiev, the Soviet geologists, to evaluate the petroleum possibilities of the Bahia, Amazon, Maranhão, and Barreirinhas basins and to present recommendations on exploration and on procedures to augment production from Bahia. Petrobrás also asked Bakirov and Tagiev to suggest how to increase the use of Russian drilling methods and equipment in Brazil. The Russians' recommendations, said Tácito Freitas, "put the lie" to everything Link had said. Yet Tagiev and Bakirov appear to have differed from Link only on the estimate of reserves, and were not positive about higher totals. They said, for example, that the "majority" of the sedimentary basins in Brazil were "highly prospective" for oil and gas, yet recommended that only a token exploration force, for "evaluation" purposes, be kept in regions outside Bahia. Further, in discussing the Recôncavo and South Tucano basins, they said only that the reserves "'*should* be more than three times" the present estimate. Their other recommendations concerning Bahia urged more efficient production methods: secondary-recovery techniques, incentive bonuses for drilling crews, improvement of services, proper preparation of fields, improvement of drilling technique, and the use of drills capable of reaching 5,000 metres.

Hugo Régis dos Reis said that the most important part of the report was the implication that Petrobrás could double its production almost immediately, if it would make the necessary investment in new equipment. He estimated that to bring about the production goals recommended by the Russians would take Cr$100 billion (about $100 million) a year. Tácito Freitas agreed that "ending the domain of the Americans within Petrobrás — who had sabotaged it flagrantly" — would not be without cost. On the basis of the report, noted Tácito Freitas, Petrobrás had immediately ordered from Rumania fifteen drilling rigs capable of reaching depths of 4,500 metres.[80]

The Russians were, in fact, much more critical of Petrobrás than published accounts indicated, particularly in the areas of exploration and production. For example, the Russians charged that the drilling Petrobrás did on its own had been to rather shallow depths, and showed up badly when compared to that done by private companies under contract. Production methods were also weak: injection of gas for secondary recovery was used too late, if at all, reflecting poor preparation of holes and a general lack of coherent programming. The Russians charged that Petrobrás was highly bureaucratized, with DEPEX and DEPRO (the production department) not co-operating. Each was overly conscious of its role, and neither felt it had to concern itself with calculations of reserves; consequently, development suffered. The Russians urged that field preparation be given to DEPEX, to give it more incentive to discover large fields. Planning appeared to be year-by-year, rather than long-range; moreover, many plans existed in different departments, and the bureaucratization of the company was such that these plans conflicted and were not examined for rationalization.[81] (Tácito Freitas, commenting on the Russians' charge, said that Petrobrás did not have enough drilling rigs because the federal Superintendency of Money and Credit, "a figurehead for the trusts", had not released sufficient hard currency for such foreign purchases.)

Walter Link, in turn, asserted that the conclusions of his report were still valid, perhaps even more so, in the light of declining production figures. He repeated his version of the geologic history of Brazil — generally unfavourable for the accumulation of oil in commercial quantities — and concluded, "Not even the Soviets could modify that situation."[82]

As 1964 opened, yet another crisis within the upper levels of its administration shook Petrobrás. On January 24 General Albino Silva called a press conference at which he accused five "high functionaries" of Petrobrás (one of whom was Director Jairo José Farias) of having negotiated "ruinous" contracts for the purchase and transportation of petroleum. Specifically, he charged that Farias had negotiated a contract with Atlas Development of Venezuela for the purchase of 500,000 barrels of oil at $1.71 f.o.b. per barrel; not only had the real cost been $2.20, but the freight had been higher than rates charged by FRONAPE, the Petrobrás fleet. Albino Silva charged that Farias had made similar "irregular" contracts with Sun Oil and British Petroleum.

Silva also attacked the superintendent of FRONAPE and the chief purchasing agent of Petrobrás.

Asked on the other hand if he himself had negotiated a purchasing contract with a foreign oil company "without hearing the Directors and disregarding reports to the contrary by technical organs", he replied that the accusation had been made by "rats" who had "resolved to sling mud" on Petrobrás itself. He appealed for the support of labour against the enemies of Petrobrás and of Goulart, who, he said, were of the same type that had driven Vargas to suicide.[83]

The *Jornal do Brasil* commented that the corruption cited by Albino Silva indicated "that the deterioration in the heart of the company is really astonishing." The picture that the president of Petrobrás painted "indicates that the decomposition is of such an order that the technical standards of Petrobrás can no longer be preserved without a sanitizing action" of a depth never before taken. The situation of the state oil company was a "national disgrace".

The five men thus indicated were fired on January 25, and immediately held their own press conference. They accused Albino Silva of trying to impose "personal solutions" on the company and of making decisions without consultation. Jairo José Farias called him a "willing agent of Standard Oil".[84]

The Petrobrás labour syndicates rose to the defence of the dismissed administrators and, in a notice in *Última Hora*, called for the removal of Albino Silva. They declared they would not accept the dismissal of any other directors or technicians, and charged that Albino Silva was merely trying to divert attention from a contract he had made with a subsidiary of Standard Oil. Régis dos Reis said that Albino Silva was attempting a *coup* because of frustration with his directors and predicted that the dismissal of Farias would "not be accepted by the workers' syndicates of the petroleum industry."

In a front-page editorial, *Última Hora* called for a "profound revision" in the structure of Petrobrás, and urged the replacement of Albino Silva by Marshal Osvino Ferreira Alves, as a "highly positive step" toward restoring confidence in the company. (Osvino Alves had commanded the First Army before his retirement; he was known as the "People's General".)[85] On

January 28 President Goulart made just that change. He also dismissed Petrobrás's three directors, Jairo José Farias, Hugo Régis dos Reis, and Alfredo de Andrade, and named an inquiry committee under Hélio de Almeida, his former Minister of Transport, to investigate the various charges.

Newspaper comments on the personalities involved in the change indicate how deeply Petrobrás had become enmeshed in politics. Osvino Alves was reputed by one newspaper to be linked with Leonel Brizola, which might mean that Goulart had lost control of the company. But the *Jornal do Brasil* disagreed, for several reasons. First, Alves had commanded the First Army and had given strong support to Goulart in 1961. Goulart was thus putting him in a position of power hoping "to pacify and discipline the Army". Moreover, the marshal had been a disciple of Estillac Leal, and had strong ties with labour. His appointment was thus an attempt to bring labour under control. Indeed, the move, rather than being toward Brizola's camp, was an attempt to wrest control of Petrobrás from the "Radical Left" and bring the company back into the patrimony of the President of the Republic. Finally, the newspaper speculated, Goulart might have made the appointment to prepare Alves as a presidential candidate, since the "People's General" appealed to all sectors of the "Left". As for Almeida, a columnist in the same newspaper suggested that his appointment was a springboard for the 1965 election, in which he would challenge Carlos Lacerda for the governorship of Guanabara state.[86]

Albino Silva was bitter about his dismissal and charged that he had been the victim of a "personal attack", which had begun the day he had assumed office. He warned of a "reign of terror" in Petrobrás, directed by men in "high office" who were trying to destroy the company, and claimed that they had already "deluded and subverted" the unions.[87]

Alves's appointment found favour among Petrobrás labour, which in an open letter to him reiterated that Albino Silva had been responsible for the crisis and had been a "pliable instrument" for groups that wanted to destroy Petrobrás. They called for "constant and frank discussions among workers, technicians and management" to implement the best policy for the growth of the state oil company. Goulart and Alves took this bit of advice to heart: they chose replacements for the dismissed executives from lists supplied by the syndicates.[88]

The inquiry committee under Hélio de Almeida convened *in camera* in the CNP library. A Congressional Inquiry Committee on Petrobrás was also in session and the *Jornal do Brasil* hoped the two groups could complement each other's work and draw up a program of "sweeping structural reform" for Petrobrás. The latter committee finally began hearing testimony about the crisis at the end of January, after having almost collapsed from disputes among its members. Cláudio Carlos Godinho, an electrical engineer with Petrobrás, charged that the Cubatão refinery was falling under Communist domination; he also attested to the "great power" that the unions had within the company. The next day Albino Silva and Hugo Régis dos Reis appeared in turn before the committee. Albino Silva said that directors linked to the unions, not "the trusts", were responsible for the crisis. Régis dos Reis, however, blamed "the trusts" for the situation, and charged that Albino Silva had been trying to set up an empire.[89]

But financial problems remained, and early in 1964 geologist Glycon de Paiva warned that Petrobrás was on the verge of collapse. He estimated that the company needed more than $21 million a month in foreign exchange ($13 million for imports of crude oil and the remainder for foreign debt obligations, contract payments, equipment and parts, and royalties on the use of processes), and that it constituted "the heaviest drain on the balance of payments of Brazil". He also charged that Petrobrás had become "the most important Communist bastion in America, after Cuba".[90]

Yet the campaign to nationalize the private refineries re-emerged, and Marshal Alves, in a February speech at Duque de Caxias refinery, said his purpose as president of Petrobrás was "to bring about the state monopoly". He added that he had a "great interest in the expropriation of the private refineries" and suggested that the decision could be announced by President Goulart at a mass rally planned for March 13.[91] This rally, in the Praça da República of Rio de Janeiro, was to mark the official beginning of Goulart's "radical-leftist" political policy. One of the two decrees he dramatically signed in front of the crowd nationalized Brazil's privately owned refineries.[92]

A week later, Petrobrás announced that the federal government would nationalize the distribution of petroleum derivatives "within the next few days" to complete the state monopoly.[93] João Goulart signed the decree on April 1 in Brasília, where he had fled in the face of an army rebellion. It was, however, one of his last acts as President. Soon afterward he retreated to Pôrto Alegre and the President of the Senate declared the office of the Presidency of the Republic vacant.[94]

The army had rebelled because of growing internal chaos and what it saw as a "drift to the left" which, if not brought on by Goulart's political ineptitude, was certainly quickened and deepened by it.[95] The chaos and radicalization that had buffeted Petrobrás was identical to that which had plagued national politics. Just as a liberal oil policy had ceased to be a political alternative prior to the creation of Petrobrás, so developmentalist nationalism lost out to radical nationalism within Petrobrás some time around 1960; for the political economy of the nation, radical nationalism increasingly smothered the more moderate variety during the Goulart administration. Strident radicals effectively silenced their more moderate colleagues within Petrobrás by characterizing even constructive critics as *entreguistas* and "agents of the trusts". As a result, the company spent the years between 1960 and 1964 trying to disprove the "Link Report", after having publicly swept it aside as unworthy of consideration. Company executives, increasingly chosen for political reasons, refused to accept the message implicitly stated in the report: that the myth of Brazil's abundant oil reserves was just that, a myth. Further, they used the other myth—that "the trusts" were eager to exploit Brazil's vast reserves — to silence critics and to further personal political goals. Petrobrás and Brazil appeared to have been falling under the dominance of demagogic politicians who used chauvinism for selfish ends; the army moved to stop the trend.

7
REASSERTION OF AIMS, 1964–1970

The 1964 *coup* was as much of a watershed for Petrobrás as it was for Brazil as a whole. Even before meaningful civilian political activity had ceased for Brazilians (in 1968) — indeed, almost immediately after the *coup* — Petrobrás had to all intents and purposes disappeared from the political arena. Before examining this most recent stage in the company's career, however, it is important to look back over its genesis and first decade to discern why it should have become so much the centre of political controversy.

One factor which kept Petrobrás too subject to political pressures was its inability to become independent of governmental sources of revenue, primarily because it had not—and has not—found enough oil within Brazil to satisfy its own needs. By Law 2004, the federal government supplied initial capital to Petrobrás, and most of its additional revenue was to come from taxes and duties. The same government was to retain a minimum of 51 per cent of company shares.[1] The Levy study of 1960 showed that nearly 60 per cent of the company's capital came from such sources, external to its operations.

Crucial to the persistent reliance by Petrobrás upon the government for revenue was the company's continuing belief that its purpose was to find oil in Brazil; its acceptance of the myths about Brazil's oil wealth and the threat "the trusts" posed continued, as we have seen, in the face of contrary assessments and recommendations in the "Link Report". Petrobrás thus continued to explore exclusively in Brazil, in effect "pouring good money after bad" and operating at a loss. It need not have done so; there are many examples of state companies—one of the best is Ente Nazionale Idrocarburi (ENI) of Italy, particularly as it operated under Enrico Mattei — competing successfully, using nationalistic rhetoric, against "the trusts" for exploration rights outside their national boundaries.[2] Petrobrás did not follow the example of such companies as ENI because the recommendation to do so came over the name of a foreigner at a time when the company and national politics were becoming increasingly populistic and nationalistic. So exploration continued in the same territory, and the public treasury continued to underwrite it.

While the company thus received substantial revenues — which it otherwise would not have had — for such important ventures as exploration, it was probably too much to hope that politicians would leave Petrobrás alone between 1957 and 1964. The company was the creature of a nationalistic campaign, which had tied it inextricably to the government in a position of obvious dependence. In the climate of the 1957–64 period, as Brazilian politics fell more and more under the sway of radical nationalism, it seems inevitable that the company would have become a "political football".

Petrobrás's dramatic advances to 1960, stirring hopes that Brazil would soon have a self-sufficient oil industry, further increased the likelihood of political meddling. The intense interest with which politicians came to view the company meant that policy decisions were subject to public, and political, scrutiny and debate. As it received larger appropriations, Petrobrás's own public-relations department tried to keep the company favourably in the public eye, which only made it more attractive to politicians.

While the rapid turnover of personnel at the executive level did not draw the company into the political arena, it must have reduced its ability to withstand such pressures. More significant was the way politics came to determine such appointments, especially after 1960.[3]

It is clear, however, that Petrobrás would not have been thrust into politics had there not been discernible widespread public support for such a move. That there was is clear, and it is attributable to two public sentiments. On the one hand, the majority of the Brazilian populace, in common with their brethren in Latin America and the rest of the third world, feared "the trusts". The reputation of the large international oil companies in this region is bad; whether such a reputation is deserved or not is irrelevant, because it is widely held to be so. Petroleum plays a large part in hemispheric and third-world development, and "the trusts" have been involved all over the region from the earliest days of the industry; hence relations with governments have been uneasy at best, and companies have often been castigated as "agents of economic imperialism". Brazilians in substantial numbers supported their state-monopoly oil company because they shared third-world fears of the "rapacity" of "the trusts". They also supported Petrobrás because they were determined to find their own oil—which they had always believed was present in great abundance — by their own efforts, since "the trusts" had earlier ignored Brazil's potential. This fear and at the same time resentment of "the trusts" not only thrust Petrobrás into the political arena—as the instrument of Brazil's resistance to the economic imperialism such entities represented —but it limited its freedom of action in dealing with them.

The second public sentiment current among the majority of Brazilians was fear of what has come to be called neocolonialism, a fear that the nation, having gained political independence, would become an economic colony of national or multi-national corporations. Horta Barbosa had articulated this sentiment in 1947 when he asserted that once "the trusts" had secured a foothold, however small, in Brazil's oil industry, they would inevitably take it over, reducing Brazil to "neocolonial" status. As later constituted, Petrobrás uniquely represented Brazil's determination to break out of its centuries-old colonial mould and become one of the world's industrial powers.

Yet the question remains, why did this postwar wave of nationalism appear, and why did it focus on oil rather than some other basic commodity? To turn to the last part of the question first, the traditional symbol of industrialization and economic independence is the steel mill, owned or controlled by the state. In Brazil's case, the enormous Volta Redonda integrated steel mill appeared in 1942, proof of the nation's move toward industrialization; not only was it a dramatic example of state planning for heavy industry, but it was a highly successful product of economic nationalism.[4] Before the end of the Second World War, then, Brazil had shown that it was serious about economic emancipation. That Volta Redonda was a success meant that the policy which produced it had been proven viable. Vargas's policy of industrialization had sired a new business elite, which had begun to compete for political power with the traditional export-oriented business sector; the industrial elite was much more nationalistic than its older opponents, having been nurtured in state initiative and control. Thus, when the postwar Dutra government attempted to relax nationalistic controls on petroleum—in the hope, it should be emphasized, of creating a viable industry—the Brazilian populace responded quickly to a call to oppose the projected move. Clearly, the postwar wave of nationalism appeared in response to the "threat" that the Dutra government's proposal posed for oil and for Brazil's industrialization.

But oil was not the only potential source of energy available; there was also electricity. Moreover, there was a postwar crisis in the supply of electricity, as well as in oil. And some Brazilians believed that the foreign-owned distributors of electricity controlled enough politicians and newspapers to stave off pressures to invest in needed new facilities; these Brazilians believed the same about foreign-owned oil distributors.[5]

Yet the crisis in electricity was gradually eased, and without the enormous public outcry that generated Petrobrás. Judith Tendler, in her excellent study *Electric Power in Brazil*, has shown that foreign electricity distributors and the state evolved a *modus vivendi* by the 1950s whereby the former busied itself principally with distribution and the latter with generation, the more glamorous sector. Although the state allowed only one rate increase between 1934 and 1965 (thereby minimizing public indignation at the distributors), it kept the distributors satisfied by granting them preferential foreign-exchange rates and allowing them to add surcharges for certain costs.

Surely the most important reason, however, for the public campaign focusing on oil rather than electricity must have been the higher level of emotion surrounding the oil industry, and the accompanying existence of much better known and apparently more powerful oil "trusts", in contrast to the situation with electricity. Moreover, as Tendler points out, there was no postwar move to roll back a nationalistic posture with electricity, as there was with oil.[6]

Thus, the postwar campaign to stop the liberalization of Brazil's oil policy caught on for four reasons. The first, mentioned above, was that the new policy appeared to relegate Brazil to its pre-1939 colonial status, and was thus particularly repugnant to the emerging industrial elite. This aspect of the campaign was still more compelling to students; on the one hand, probably because it appeared to promise better employment opportunities, and on the other, because many orators appealed to students to lead a willing populace against a treacherous government. Second, much of the campaign's attraction arose from the political release after the Vargas dictatorship: not only was there a residue of nationalism among the masses (Vargas had used nationalism to justify his rule), but the appearance of democracy suddenly added to this legacy a sense of participation in the national destiny. The oil issue was a superb focus — the first to appear after Vargas's removal — for this new awareness of national duty. Third, Brazil had discovered oil under government aegis in 1939, despite assertions from "the trusts" that it had none. Why should it let those feared groups in now? Finally, the proposal to liberalize oil policy fell victim to political factionalism, as Vargas campaigned for almost certain election in 1950 on a platform of continued nationalism to stimulate Brazilian growth.

With Vargas elected in 1950 and the "anti-national" oil proposal clearly repudiated, the campaign continued, but to ensure that Vargas's Petrobrás would not allow foreign participation in the industry. With ultimate success, economic nationalism was further vindicated. If the anti-Vargas forces, found chiefly in the UDN, had come to support the legislation creating Petrobrás because they believed, or at least hoped, that the company would fail, its success made their policy backfire. Liberalism was thus driven further into political limbo.

Vargas's subsequent suicide in 1954, and the discovery of his inflammatory letter castigating anti-national forces, consecrated economic nationalism for much of the Brazilian public. It also served to render Petrobrás "untouchable" in the eyes of its defenders. Thereafter (particularly after 1957), these defenders — increasingly radical nationalists — acted as if they owned Petrobrás. While they took it upon themselves to repel all real or imagined attacks on the company, they were also not averse to using the company's successes for their own ends, urging it on to new triumphs. Petrobrás was thus under strong pressure to produce, and responded by increasingly trying to keep itself in the public eye. Thus, every "show" of oil was trumpeted as a discovery, and when production began to stagnate, Petrobrás's publicity turned to an emphasis on company "public works" as benefits to the nation. The task set out for Petrobrás — the creation of a self-sufficient, integrated, national oil industry — was difficult enough; having to make decisions in the glare of political struggle hindered the company and almost weakened it irreparably as it strove toward its goal.

Early in April 1964, following the consolidation of the army revolt, Acting President Ranieri Mazzilli offered the presidency of Petrobrás to General Olímpio Maurão Filho, the first conspirator to mobilize his troops.[7] General Maurão refused, saying, "Of petroleum, I hardly know the smell."[8] On April 7, General Ademar de Queirós took over, charged with the "total cleansing of extremist elements" from the company.[9]

The army moved swiftly to neutralize the resistance it expected from Petrobrás's employees. Troops found eleven machine guns at Duque de Caxias refinery, and a gun-battle between employees of opposing political views was threatened during the first week of the revolt. A Petrobrás tanker, the *Ipojuca*, had sailed for Rio Grande do Sul with fuel to help Goulart in his anticipated final stand. Union leaders called a strike of all oil workers, but the army stopped it. The police of Guanabara uncovered a plot to add water to the gasoline produced in the Duque de Caxias refinery, prompting the Minister of War to order the entire staff of Petrobrás in the Rio de Janeiro area imprisoned. A committee immediately began a thorough investigation of the company's operations, with special attention to the "fabulous expenditures" in publicity.

Brazilians assumed that the new political order would act ruthlessly, and General Queirós immediately dismissed seventeen employees at the level of chief; journalists believed that a mass purge of Petrográs personnel would soon follow — especially since the new chief of the Petrobrás Public Relations Division, Brigadier Ari Neves, said that there had been financial irregularities in the division before the revolt.

The initial move by the army toward cleansing national politics was to arrest the most notorious radical nationalists and deprive them of their political rights for ten years (meaning that not only could they not vote, but they could neither participate in politics nor comment publicly on political matters, nor could they hold positions with public or autarchical companies). Among the first so penalized were former Petrobrás presidents Francisco Mangabeira and Osvino Alves, and company directors Hugo Régis dos Reis and Jairo José Farias (the latter exchanged shots with police before submitting to arrest). Mário Lima, who had led the Mataripe refinery labour unions before becoming a federal deputy, called on workers to blow up the refinery in retaliation for his arrest and loss of political rights. No sabotage, however, was reported.

By mid-April Petrobrás had begun to return to normal operations, signified by an announcement from the company directorate that the exploration effort would be intensified. To this end, eighty-two new teams would be sent into the field, concentrating on the states of Bahia, Sergipe, and Maranhão. Coincidentally, the new Brazilian President, General Humberto Castello Branco (who had led the *coup* which toppled Goulart), announced that he would "reformulate" the national policy through the CNP, as soon as he had appointed the agency's president.[10]

The Petrobrás directorate also announced that it would soon re-examine the nationalization of private refineries and distribution which Goulart had decreed just prior to the *coup*. The former owners of Capuava refinery had already challenged the legality of the decree, and it was quickly referred to the courts. Although Petrobrás fought to have the decree upheld (but not too hard — the company had had enough publicity), it was ultimately quashed by a higher court and the National Security Council. The refineries in question resumed operation as private entities.[11] Apparently, no further mention was made of the nationalization of distribution, and the decree was not implemented.

In May, in an effort to force Petrobrás to become more efficient as well as to curb inflation, the government ended the exchange subsidies on imports of petroleum. The retail price of gasoline immediately doubled, and it was expected that the cost of public transportation would rise accordingly. The subsidy on imports of crude oil did not, however, disappear entirely; the Bank of Brazil was permitted to maintain special prices below the free market value of the dollar for imports of fuel oil and LPG. The *New York Times* said that the exchange subsidies on various commodities had been costing the government $200 million per year, and had been a major source of inflation, since they had been covered by issuing currency. When asked about the extent of the rise in gasoline prices after the government had announced the curtailment of subsidies, Ademar de Queirós said that he had no statement and would follow the orders of the President of the Republic (rather a contrast to previous executives of the company!).[12]

Throughout the early months of the new administration, Petrobrás was under examination by a number of investigating committees. They deliberated in secret and it soon became clear that the government and the executive of Petrobrás intended to withdraw the company from the "public eye". On May 23 the *Jornal do Brasil* reported that according to the public relations division of Petrobrás the various investigating committees had not yet called for the dismissal of any employees. Proceedings were so secret that there was no announcement to the public about the number of such committees actually at work probing into the past of the company (there may have been as many as thirty-one). The military police were conducting an investigation, as was the company itself. The last information about financial scandals in the company was released on April 24, 1964, and concerned price cuts given to various organizations within the armed forces, student and syndicalist groups, and certain divisions of the military police.[13]

In June the committees announced that they had completed their work; rumour had it that 2,000 of the 40,000 Petrobrás employees were suspected of subversive activities.[14] The superintendent of the executive police, Colonel Livio Galdo, exhibited documents concerning misappropriation of substantial funds by the public relations division of Petrobrás; the money had been used for "subversive" purposes, he said, citing expenses incurred for banquets and parties, hospitality in "luxurious hotels", trips, publicity, and propaganda. He also charged that the company magazine, *Petrobrás*, had been used "to expound the Communist ideology" within the oil industry.[15]

By mid-July only sixty-one employees, fifty-one of whom had been involved in the brief 1963 strike to force the expropiration of Capuava refinery, had been dismissed. The complete findings of the various investigating committees were never divulged, and no more employees lost their jobs as a result of the inquiries. That so few employees suffered as a result of the inquiries, and that the proceedings of the inquiries were suppressed, was due to General Ademar de Queirós, then president of Petrobrás. He resisted what was at times enormous pressure to clean out whole departments, and agreed to dismiss employees only in the most extreme cases of incapacity. In the opinion of oil pioneer Irnack Carvalho do Amaral, Queirós saved Petrobrás from at least sharp curtailment of its activities if not actual destruction.[16]

In June 1964 President Castello Branco signed a decree regulating petroleum imports which attempted to ensure that Petrobrás would get the best possible crude-oil and derivatives prices for its own, and the privately owned, refineries. By the decree, Petrobrás could not sign with any one producer for more than a twelve-month period or for more than 10 million barrels of crude; contracts for derivatives could not run for more than six months, nor could they be for more than the national consumption of such commodities during the same period. As an incentive to drive for the best price, Petrobrás was allowed to keep 50 per cent of the profits from imports of crude and derivatives; the amount was to be used to meet its financial obligations abroad and to purchase equipment in foreign countries. The decree also laid down that the foreign purchase of crude oil and derivatives was to stimulate the Brazilian economy: all such contracts had to provide for the purchase, by the supplier, of Brazilian products to the value of at least 20 per cent of the contract price. Such products were to be chosen by the Committee of Foreign Commerce, to forestall any possible conflict of interest on the part of Petrobrás.[17]

Not to belabour the obvious, the 1964 revolt had a profound effect on the political activities of Petrobrás. Just how profound it was to be was indicated in a speech President Castello Branco made in August 1964 to workers at the Duque de Caxias refinery. He said that the company would henceforth remain above party politics and ideological interests, and concern itself solely with the technical aspects of the oil industry.[18] Thus, "completion of the state monopoly" ceased publicly to interest Petrobrás, and it appeared that the company's activities would not exceed those enumerated in Law 2004. In less than a year the administration demonstrated that Petrobrás was not to be allowed to extend its control over the national oil industry. In July 1965 the National Security Council approved private investment, of any origin, in the petrochemicals industry. Soon after, the council opened development of the southern oil schists, hitherto reserved to Petrobrás, to private companies formed in Brazil. It should be noted that the company has been reduced to actual competition with private enterprise in only one of these two areas, petrochemicals; it retained a preferred position in schist development. The decree concerning the latter reserved the most promising schist deposits, around São Mateus, Paraná, to Petrobrás; private companies developing other deposits had to sell the oil or gas produced to Petrobrás at prices no higher than those on imported equivalents. (Until today, this would hardly have been called an attractive investment; only since the massive rise in crude prices early in 1974 can oil distilled from shale be produced at a price competitive with the current world crude price.)

Evidence of the world political orientation of the Castello Branco government came in the same decree, since Petrobrás was required to break off negotiations with the U.S.S.R. over exploitation of the schists. Petrobrás then signed a contract with the Cameron & Jones Company, of Denver, Colorado, to build a pilot distillery at São Mateus.[19] The plant continues to operate, but has not (by 1974) gone beyond its initial stage, presumably because production is still too costly. The schists are now being looked at for their natural-gas reserves, the raw material upon which Brazil's booming petrochemicals industry is based.

Petrochemicals is to be an industry largely in private hands, with Petrobrás participating on a limited basis through a subsidiary. Petrobrás does not wholly own this subsidiary, Petrobrás Química (Petroquisa); 49 per cent of its shares were reserved for private interests. True, for the immediate future all the petrochemicals companies in Brazil, whether Petroquisa has an interest in them or not, must buy their raw materials from Petrobrás (chiefly naphtha and natural gas), but Petroquisa does not appear to have a privileged position in the industry.[20]

Exploration activity increased substantially after the 1964 *coup*, followed by gratifying discoveries and a steady growth in production. In December 1964 Petrobrás announced the discovery of a field near Carmópolis, Sergipe, in a region for which the "Link Report" had recommended further exploration. Initial estimates of its potential yield ran from 3,000 to 25,000 b.p.d., but by mid-December 1966 production had risen to only 10,000 b.p.d.

The reasons for the discrepancy and slow rise in production are several. The variation in production estimates arises principally from political factors: the new national and company administrations wanted to demonstrate success, and so issued generous figures. (When a field is discovered its potential yield is a matter for conjecture until the field's limits have been found.) There may also have been vestiges of the pre-1964 over-enthusiasm re-surfacing briefly within Petrobrás.

The lag in production arose from a natural factor: Carmópolis oil is even thicker and heavier than that found in the Bahia Recôncavo, and thus even more difficult and costly to handle. Its best use is for asphalt, for which Bahia oil is also well suited. Having no urgent need for such oil, Petrobrás delayed building storage facilities and a marine terminal until the end of 1966, at which time production rose to 10,000 b.p.d.[21]

In the subsequent four years, Petrobrás discovered other fields in Sergipe and neighbouring Alagoas, which helped to raise production from the region to 30,000 b.p.d. by the end of 1970. Exploration of the continental shelf adjacent to Sergipe began before the end of the decade; several promising "shows" were soon obtained, prompting optimistic predictions about the zone's potential from Petrobrás (the optimism was expressed in more guarded terms than had been customary prior to the *coup*).[22] In 1966, Petrobras discovered the Miranga field in the Recôncavo, which produced almost 40,000 b.p.d. by the year's end.[23] In the same year it brought in a producing well in the Barreirinhas basin on the coast of Maranhão state (east of São Luís). On December 30, 1966, Petrobrás president Irnack Carvalho do Amaral* said the event signified "the discovery of a new oil province in Brazil". He was optimistic that Petrobrás would find a commercial field in Barreirinhas with further drilling. Unfortunately, Amaral's optimism has so far proved groundless; not only was Barreirinhas found not to be a commercial field, but prospects in the region have declined as exploration progressed.[24]

For some time Petrobrás showed substantial gains in production under the new political order. Production began to rise steadily late in 1965, and in 1966 daily averages increased by greater and greater amounts. On December 27, 1966, President Irnack Carvalho do Amaral announced that totals had passed 150,000 b.p.d., from 102,000 b.p.d. at the first of the year.[25] Thereafter the pace slowed. In 1969 production had increased to only 172,000 b.p.d., and in 1970 it actually declined to 164,000 b.p.d. The decline was in the Recôncavo — apparently the start of a trend — and a marginal increase from fields in Sergipe-Alagoas could not offset it.

By 1970, reserves were at 857 million bbl., an increase of one-third in the decade since Walter Link had left Petrobrás. The rate of increase, however, had slowed here as well: the 1970 figure was only 0.6 per cent higher than that for 1969.[26] In other words, at that time Petrobrás was doing little more than replacing depleted fields with newly discovered ones. Petrobrás has also for some time used secondary-recovery techniques (chiefly the injection of natural gas into wells to force out remaining oil after the well's own pressure has been exhausted) to boost

---

* President Castello Branco made Marshal Ademar de Queirós Minister of War in July 1966 in response to a political crisis. Amaral served as Petrobrás's president — his second term — from July 4, 1966, to April 4, 1967.

production figures. Such a practice, while costly, is common in oil production; Brazil's unfortunate petroleum geology (weak rock structures, meaning little internal pressure in oilfields to drive the oil to the surface) definitely calls for widespread use of this technique.

In 1969 Petrobrás embarked on a policy of becoming self-sufficient in oil by the end of the following decade (not a novel declaration — Petrobrás has periodically set such deadlines in the past), clearly placing its hope in the continental shelf. The company is to explore about 800,000 square kilometres of the shelf, mainly adjacent to the Amazon delta and the states of Sergipe, Espírito Santo, and Rio Grande do Sul. By the end of 1970 two wells off Sergipe had shown evidence of substantial amounts of oil.[27]

Schist development, which would contribute to production totals, remained questionable at the end of the decade. The pilot plant at São Mateus, Paraná, was then distilling 1,000 b.p.d. of oil from shale, but at a price that was obviously prohibitive at that time. In 1966 the Ministry of Mines and Energy reported that "competent geologists" considered the recoverable reserves to be 300 billion barrels, 400 times the national reserves recoverable by drilling. The Ministry said, however, that Brazil still faced technical problems in the development; it did not reveal what the problems might have been.[28]

Petrobrás publicity after the 1964 *coup* thus continued to defer to the two national assumptions — that Brazil is rich in petroleum and that "the trusts" are eager to exploit it — upon which the company is built. Probably because of the pessimistic outlook for production, however, it has switched its emphasis from the exploration and exploitation of crude oil to other aspects of the industry. One can argue convincingly that so long as there is no return to the political style of the Goulart era this continued catering to national myths is producing few adverse effects beyond the expending of money in the continued exploration of the national subsoil.

In the refining sector Petrobrás and the private refineries have produced an increasing share of the derivatives consumed in Brazil since 1962. By 1970 only 2 per cent of products consumed in Brazil were imported; self-sufficiency in this area is very close.²⁹ There were five Petrobrás refineries by 1970: Duque de Caxias (150,000 b.p.d.), Cubatão (115,000 b.p.d.), Mataripe (64,000 b.p.d.), Canoas, Rio Grande do Sul, and Betím, Minas Gerais (each 45,000 b.p.d.). A sixth, Paulínia, São Paulo (126,000 b.p.d.), was under construction with completion in 1972. Both Duque de Caxias and Cubatão were to be expanded 50,000 b.p.d. by the same time.³⁰ Petrobrás is committed to achieving and maintaining self-sufficiency in refining capacity and refined products, and there is no reason to doubt that it will achieve its goal.³¹

The state oil industry has played a role in the development of Brazilian higher education. In 1951 the CNP signed an agreement with the University of Brazil to begin training petroleum technicians. The first course—for refinery technicians—began July 1, 1952. Now several Brazilian universities are offering a variety of geological and technological programs, and most Petrobrás personnel have been trained within the country.

Petrobrás stimulates domestic industry by buying as much equipment as it can within the country, a percentage that has steadily increased. Petrobrás has also given national industry an export market: in recent years 20 per cent of imported petroleum has been paid for with Brazilian exports. Around 1970, studies were being made of means to raise the proportion to 40 per cent.³²

Since 1961, when Petrobrás opened it first service station in Brasília, the company has become a factor in its own right in the distribution sector. By 1965 there were 89 Petrobrás outlets; by the end of the next year their number had almost doubled, to 174; at the end of 1970 there were 527 Petrobrás service stations. In 1966 Petrobrás was third of the six distributing companies (behind Shell and Esso) in terms of sales, controlling 13 per cent of the national market; in 1970 it was in the same position with about 20 per cent of the market.³³

In recent years Petrobrás has publicly asserted that it has no intention of monopolizing distribution, and that it is content to share the growing market with Shell, Esso, Atlantic, Texaco, and Ypiranga (a private Brazilian distributing company) and to compete with them. For a time after the 1964 *coup*, however, lessees of privately owned service stations believed that nationalization in some form would surely come, and a number of them switched their leases to Petrobrás, in recognition of what they saw as the inevitable. This movement appeared to have died down by 1970.

Petrobrás is also steadily adding to its fleet, to save foreign exchange (in the long run) by internalizing tanker charges. It is also effecting short-term exchange savings by purchasing a growing proportion of its ships from national yards. (The tonnage built in Brazil is also increasing as larger and larger ships are being built.) By the end of 1970 the Frota Nacional de Petroleiros (FRONAPE) comprised thirty-two ships, with a deadweight tonnage of 820,000. Orders had been placed for several more ships with yards in Yugoslavia and Japan, among which was a 265,000 dwt oil and bulk carrier.[34]

Petrobrás has become a giant among the world's companies. In 1966 Petrobrás made its first appearance in *Fortune*, in its annual review of the "200 largest industrials outside the U.S. in terms of sales", in eighty-eighth position. The magazine listed Petrobrás's sales at $527,958,000, assets at $550,992,000, and net profit at $87,360,000. For 1970, Petrobrás was again in eighty-eighth place; sales were $875,212,000, assets $1,493,038,000, and net profit $164,072,000.[35] These figures, intriguing in themselves, must be viewed with some caution. For example, an inflationary rate of over 25 per cent annually in the latter half of the 1960s reduces the profit figure. Then, too, the federal government occasionally granted special subsidies to Petrobrás; for example, when he became president of Petrobrás in April 1967, General Artur Candal Fonseca asked the government for more funds, and the new Minister of Mines and Energy, Colonel Costa Cavalcanti, promised money to increase national crude-oil production.[36]

With the above incident in mind, one might suspect that many of the conclusions that the New York firm of petroleum consultants, Walter J. Levy, Inc., drew in 1959 about Petrobrás's sources of funds would still be valid. While an exact comparison cannot be made (Levy was dealing with a six-year period, and some categories have changed on the balance sheet), it might be illuminating to view Petrobrás ten years later, according to the 1969 shareholders' report.

Table 2

Petrobrás: Sources of Funds, 1969[37]

	MILLIONS OF NEW CRUZEIROS*	PER CENT OF TOTAL
Internal	524	27
Retained earnings	369	
Depreciation and depletion	155	
External	1,393	73
Increase in share capital	278	15
Receipts from sole tax on lubricants	1,115	58
Total Funds	1,917	100

* Multiply New Cruzeiros by 1000 to arrive at cruzeiros; the devaluation occurred in 1967. The NCr$ was worth U.S. $0.24 in 1969.

The most striking feature of this breakdown in comparison with that for the period 1953–59 is the even heavier reliance by Petrobrás on external sources of funds: 73 per cent as opposed to 54 per cent. The largest single source of working capital was the proceeds of the sole tax on lubricants, all of which still went to Petrobrás. This indicates an enormous volume of sales of the company's products, plus increased underwriting of the company's operations by consumers of petroleum derivatives. Sales for 1969 were NCr$2,866,900,000, a figure which must be tempered by an inflationary rate of over 20 per cent. Unfortunately, no assessment of product prices could be made; in its report, Petrobrás claimed prices had been reduced to levels "comparable to other countries", but whether or not the CNP still maintained them at higher than market level could not be ascertained.[38]

In 1969 Petrobrás was still in the "pioneer" stage of development; indeed, it seemed even further from self-sustainment than a decade earlier. To put matters simply, it had not found enough oil to pay its expenses. The importance placed on the search for oil may be seen in the percentage of investment that exploration and development received (slightly over 50), as opposed to refining (16), petrochemicals (3), transport (23), shale processing (2), distribution (2), and miscellaneous (slightly over 3).[39]

In August 1965 Professor José Eugênio de Macedo Soares declared that the Brazilian petroleum industry had progressed beyond the "pioneer phase, of political processes, of emotive solutions, in the search for objectives" and had "reached maturity". Brazilian industries, he said, had become "entirely dedicated to production and to the improvement of their activities in a climate of calm, of respect and of security."[40]

An incident late in 1967 showed, however, that Petrobrás personnel had not forgotten the rhetoric used in the past to deflect criticism. In October the *Jornal do Brasil* summarized a report on national problems from the Federation of Commercial Associations of Brazil. For petroleum, the report showed that, at the current production rate of 150,000 b.p.d., Brazil would exhaust its known recoverable reserves of 688,000,000 barrels by 1979. Should production keep pace at 40 per cent of consumption, reserves would be exhaused by 1977. The report also questioned whether Petrobrás could attain self-sufficiency in crude-oil production by 1975 (as the Ministry of Mines and Energy had predicted in 1966): in 1965 Brazil consumed 117,000,000 barrels of oil (an average of 320,000 b.p.d.); the federation estimated that consumption would rise 8 per cent per year to 253,000,000 barrels in 1975 (700,000 b.p.d.); self-sufficiency by 1975 would mean that Brazil would require reserves of 4,200,000,000 barrels, and that Petrobrás would have to discover 3,512,000,000 barrels of oil in ten years.

Two days later the same newspaper printed the reply of the "Petrobrás technicians", who charged that the report was "technically faulty", and that its object was "to induce Petrobrás to minimize investments in the refining sector". In a regrettable lapse into the polemics of "O petróleo é nosso", the technicians pointed to "internal and external pressures" linked to "various foreign economic groups" who apparently had indicated that they would not invest in the Brazilian petrochemicals industry without control of refining. The attacks against Petrobrás, continued the reply, were "now more sophisticated", based on the theory that if Petrobrás were to discover in the minimum time the necessary reserves for national consumption, the company would have to invest "massively" in exploration and thereby decapitalize itself. The technicians closed by charging that some groups were trying to return to the era of Walter Link, wasting $300 million on "dry holes in the Amazon Region".[41]

Clearly, Petrobrás officials still regarded comment or criticism as a form of corporate insult, to which they had to respond in kind.

The year 1970, the end of this particular narrative, marked a turning-point for Petrobrás, since it was the year in which the company began to explore in other countries. The agency to be used was a subsidiary company, Petrobrás Internacional S.A. — Braspetro, wholly owned by Petrobrás.

Although Petrobrás did play an important role in negotiations with Bolivia in the late 1950s, and did insist on participating in — if not controlling or even monopolizing — exploration of the Bolivian treaty zone, it is doubtful that Petrobrás would actually have explored there at that time. Its policy was clear: to find Brazil's oil. The Bolivian episode was thrust on Petrobrás and does not signify, in my opinion, a willingness on the part of the company to explore outside Brazil prior to 1970.

The years since 1970 have seen Braspetro complete exploration contracts with several countries, both in the Arab world and in South America. Until recently, however, Petrobrás has insisted that such contracts do not signify an admission that Brazil does not have abundant oil; rather, the company has said, they are for national security, to allow it to maintain Brazil's oil in reserve (a sound idea, if a little late in coming).[42] But despite such publicity, it appears that by 1970 Petrobrás executives had accepted that the myth that Brazil's subsoil was saturated with oil was just that: a myth. The likelihood of finding a bonanza field has diminished as exploration has proceeded (with the possible exception of the continental shelf, still in the pioneer stage of exploration); production from Brazil had stagnated and even declined by 1970, and totals had not increased significantly by 1974. This disappointing result can be attributed to Brazilian geology which, to put it simply, does not favour the accumulation of petroleum in fields of substantial size.[43] That was the conclusion of the "Link Report"; ten years after its presentation, Petrobrás had finally accepted both its assessment of Brazil's petroleum geology and its recommendation to explore outside the country.

The Brazilian oil industry came into being because of government stimulus, not through private initiative. Meaningful exploration began when the national government took control of it; the result was the discovery of oil. That the search for this energy source, so vital to Brazil's quest for modernity, was undertaken at all was due to a political decision, responding to two beliefs that were, and doubtless still are, current in Brazil: that the national subsoil was rich in oil, and that "the trusts" were eager to exploit it. (It bears repeating that "exploit" could have meant either that "the trusts" would extract all of Brazil's oil and sell it where they wished, or that they would hold Brazil's oilfields in reserve. In either case, what mattered was that foreign companies would "exploit" a national resource according to *their* needs, not those of Brazil.) There followed, after an attempt to forestall it, the creation of a virtual state-monopoly company, to fashion and control an industry worthy of Brazil's aspirations. For a complex of reasons then — xenophobia, corporatist beliefs in national security, national pride of varying degrees of radicalism, anti-colonialism, and political developments — the federal government took upon itself the dominant role in creating a national oil industry that would find and process national reserves of crude petroleum.

But the popular assumptions that prompted the Brazilian government to create a national oil industry have been proven wrong. Brazil does not hold abundant oil reserves, and foreign oil companies have not shown much interest in exploration, except when supplies from elsewhere were jeopardized by war. Brazil's possible oil reserves appear to have attracted "the trusts" only as a potential source during emergencies. Thus, the assumptions are no more than myths.

This book has shown how the myths came to be accepted, to

become part of Brazilian nationalism. It has also shown how Petrobrás, to all intents and purposes a state-monopoly company, came to the fore as apparently the only solution to the problem indicated by the myths: how to develop Brazil's reputedly vast oilfields without having "the trusts" seize control of them.

Indeed, from an entirely realistic standpoint, Petrobrás *was* the only viable solution, even if the problem were only to carry out a thorough exploration of Brazil's sedimentary basins. Foreign companies had not, with few exceptions, indicated any interest. Brazilian private enterprise had demonstrated outstanding incompetence in this activity, and was clearly unequal to either task. As a vertically integrated company, Petrobrás took control of the areas of exploration, production, refining, and transportation, and has come to influence marketing strongly. Despite mismanagement (notably prior to 1964), Petrobrás has operated effectively in those sectors; indeed, in such realms as exploration and production, it is clear that Petrobrás was the only entity that would have carried out work in Brazil. The results of the company's operations are: 1) economies of scale, 2) long-term savings of foreign exchange, and 3) stimulus to national industry, by both supplying energy and providing markets for manufactured goods in its own operations and through payment policies for crude oil purchased elsewhere.

But, as this book has demonstrated, there have been several things wrong with Petrobrás and the policy that created and directed it. First and foremost, politicians and company executives uncritically accepted the two national myths about oil as overriding guides for policy. As a result, much money has unquestionably been wasted, a wastage that went on for ten years after presentation of the "Link Report". All too many politicians and company executives irresponsibly lent credence to those myths, most notably between 1960 and 1964. Finally, Petrobrás refused to begin exploration outside Brazil in territories with geology more favourable to the accumulation of oil, particularly in fields of "bonanza" proportions. Such a move, if taken immediately after Walter Link and his colleagues had recommended it in 1960, would probably have made Petrobrás self-sufficient in production by the end of the decade and improved its financial position immeasurably; the move was not made during the period examined in this book because company executives allowed the national myths about oil to dictate policy. The result of these factors was that Petrobrás's operations had to be underwritten by the Brazilian taxpayer to an increasing degree in the period under study. From a financial standpoint, then, it is at least questionable whether Petrobrás can be justified.

But an examination of Petrobrás cannot be made solely or even principally from the point of view of finances; political and even emotional factors must be taken into account, and they may well be crucial — at least as far as the Brazilian populace is concerned. These political and emotional factors may be lumped together in one word: nationalism. The development of an oil industry was very important to Brazil's national pride, and Petrobrás, the instrument to effect that end, emerged as a virtual state-monopoly company because of strong mass sentiment in favour of that form. Its achievements — real or imagined — since its creation have stimulated Brazilian nationalism. Such a sentiment can be viewed positively because it fostered national development, and development, in turn, is of strategic benefit to Brazil in providing for a better life for future generations as well as the maintenance of Brazilian independence. How does one enter such considerations into a balance sheet?

The story of the development of Brazil's oil industry has illuminated four different types of nationalism. First there emerged a xenophobic or chauvinistic variety, an instinctive nativism that manifested itself simply in antipathy to foreigners. It was present into the 1920s. After the 1930 revolution came corporatism. Its main thrust was security, and its interest in the *development* of resources distinguishes it from the earlier variety of nationalism. After the Second World War, the two varieties merged to defeat the Petroleum Statute, and then a new strain emerged: developmentalist nationalism. It was primarily concerned with development, and nationalism was to be ensured by controls; there was a place for private capital, national and even foreign, in the thinking of its proponents (who tended to be from the new industrial business elite). At about the same time, however, radical nationalism appeared, which more closely resembled chauvinism than either of the other varieties. It catered to fear and suspicion, and with respect to oil played on the two popular myths. By 1960 its proponents had effectively silenced the more moderate developmentalist nationalists. It is regrettable that, despite the positive aspects of Brazilian nationalism as it was identified with Petrobrás, it was radical rather than developmentalist nationalists who ultimately came to dictate policy.

The military played a crucial role in the development of Brazil's oil industry. Since the army had been instrumental in the success of the 1930 and 1937 *coups*, Vargas had to defer to the wishes of influential officers; their concern for national security and the protection of energy supplies led the dictator to establish a controlling agency for oil, the CNP. After the Second World War, officers participated actively in the debates over the Petroleum Statute, articulating positions and leading pressure groups that represented the two principal alternatives then available: corporatistic state monopoly and controlled liberalism. During the subsequent debates over Petrobrás, officers continued their active participation, performing important roles for the two factions (supporting either state monopoly or Petrobrás) both in and out of Congress. With the triumph of a nativistic Petrobrás, the military pledged it unflagging support; thereafter, in company with the civilian political spectrum, the military split into developmentalist and radical factions. The 1964 *coup* saw the military's commitment to Petrobrás reaffirmed as an instrument to develop the nation's industrial capacity in its drive toward economic independence and great-power status. Throughout this process, the military has not so much influenced as reflected and articulated public opinion in the national exercise of creating an oil industry. Officers have acted within rather than upon the political system, responding to, rather than attempting to shape, public opinion.[44]

To reiterate, Petrobrás came into being as a vital symbol of the massive drive toward industrialization that acquired momentum after 1930, whose form was shaped by nationalism and scarcity of capital. Petrobrás, a product of that drive, emerged as the culmination of two universally accepted myths: that Brazil was rich in oil, and that foreign companies craved it. Petrobrás was unable or unwilling to move beyond the myths until 1970;

the result, to that time, was that the company's effectiveness was occasionally threatened, sometimes to a serious degree, and it was squandering money in a fruitless quest. At the same time, however, the national oil industry that Petrobrás created has brought many benefits to Brazil, benefits that will become more important now that the company has risen above the myths and begun to look outside the national territory for an assured supply of oil.

POSTSCRIPT: SELF-SUFFICIENCY BY 1979?

At the end of November 1974 Petrobrás announced the discovery of a number of oilfields on Brazil's continental shelf, which would raise production by three to five times within five years. The news was dramatic: production had stagnated at less than 200,000 b.p.d. for the previous few years, while consumption had risen steadily to four times that amount.

Another encouraging aspect of the announcement was its timing: the government did not authorize its release until after the national elections. Clearly, Petrobrás was deliberately kept out of the campaign, even though the government party knew it faced electoral losses.

Newspapers and politicians immediately predicted self-sufficiency in crude-oil production by 1979. Such predictions have appeared before; this time, however, granting three factors, it may come true. First, the rate of increase in consumption must decline. To date, there has been little noticeable effort to that end, certainly not compared to the drive in Europe and North America. Moreover, the government places heavy reliance on the motor-vehicle industry to stimulate the economy. But if future increases are, as has been predicted, in the form of public-transport vehicles rather than cars, fuel could be conserved without a check on the economy. The second factor is that Brazil must go on trading its oil on a one-for-two basis with the U.S. This means that Brazilian vehicles must continue to burn fuel containing a high percentage of sulphur, a major air pollutant. It is ironic that Brazil's high-paraffin crude, which was regarded as a liability until recently, has suddenly become very valuable: it contains almost no sulphur. The United States, concerned about air pollution, is willing to give two barrels of high-sulphur crude for one of the "cleaner" variety. If this situation and the respective attitudes toward pollution persist— and there has been no evidence to date, in either country, that they will not — the amount of oil available for consumption in Brazil will be considerably higher than national production. Finally, the world oil price level must remain high, which will allow exploitation of sources — the schists, for example — or fields previously considered uneconomical. Granting these three factors, then, Brazil may well reach self-sufficiency in crude-oil "production", as a result of explorations in her own territory, by Petrobrás's twenty-fifth anniversary.

APPENDIX I

Decree-Law No. 538, July 7, 1938

The President of the Republic, having consulted the Federal Council of Foreign Commerce, considering the disposition of Decree-Law No.395, April 29, 1938, and using the prerogative conferred upon him by Art. 180 of the Federal Constitution,
Decrees:

Art. 1. The National Petroleum Council, created by Art. 4 of Decree-Law No.395, April 29, 1938, autonomous, directly subordinate to the President of the Republic, is composed of a President and eight (8) Councillors, all named by decree.

The Councillors shall be:

1 — a representative of the Ministry of War;
2 — a representative of the Ministry of the Navy;
3 — a representative of the Ministry of Finance;
4 — a representative of the Ministry of Agriculture;
5 — a representative of the Ministry of Transport and Public Works;
6 — a representative of the Ministry of Labour, Industry and Commerce;
7 — a representative of the industrial organizations;
8 — a representative of the commercial organizations.

Art. 2. For the President or a member of the National Petroleum Council it is necessary:

a) to be a native Brazilian, of well-known competence and spotless reputation, and more than thirty years of age;
b) to be in the possession of his civil and political rights;
c) not to have at the moment of designation, nor to have had in the five preceding years, direct or indirect interest in private companies, which are dedicated or have been dedicated to the exploration, exploitation, industrialization or commerce of petroleum and its subproducts.

Art. 3. The Councillors, representatives of the Ministries, shall be chosen from among the general or superior officers on active service, civil servants of high rank, teachers in public schools or persons outside public service; the representatives of organizations shall be chosen from lists in triplicate, one for industry, the other for commerce, made, respectively, by the Industrial Confederation of Brazil and by the Federation of the Commercial Associations of Brazil.

Sole Paragraph. The President and the members of the National Petroleum Council, holders of the confidence of the President of the Republic, are invested in the form of a committee, and may be substituted or redirected.

Art. 4. The Council shall have a Vice-President designated by decree from among the Councillors.

Art. 5. The President, the Vice-President, and a Councillor designated in the manner of the above article, shall constitute the Executive Committee of the Council.

Sole Paragraph. It is prohibited to the members of the Executive Committee, while they serve on it, the exercise of any function, duty, or position in public administration, being, however, assured to the public servant, civil or military, in the exercise of the new function, the rights and privileges which are due to them on regular or active service, excepting the remuneration.

Art. 6. The members of the Executive Committee shall have their salary fixed by decree; the remainder shall receive a daily stipend for each session which they attend, fixed in the same manner.

Art. 7. The President of the Republic, through the proposal of the National Petroleum Council, shall create by decree the technical and administrative organs necessary for the service of the Council, with the respective staff, salaries and fees.

Sec. 1. The staffing of these offices shall be, whenever convenient, and in the light of need, preferably by the transfer of technical and administrative employees attached to the diverse offices of the Civil Service.

Sec. 2. The Council shall draw up its internal regulations, and shall submit the scheme for the approval of the President of the Republic.

Sec. 3. The technical and administrative organs, to which this article refers, shall be grouped into three sections, each one directly subordinated to one of the members of the Executive Committee, the President exercising general supervision.

Art. 8. The National Petroleum Council will meet once a week and, extraordinarily, whenever the President may call it, the majority of Councillors attending.

Sec. 1. The decisions of the Council shall be made by a majority of votes, the President able, through his own votes, to break a tie.

Sec. 2. To the Ministries of War and of the Navy, singly or in concert, will fall the right to protest, without declaring a reason, and with suspensive effect, to the President of the Republic, any decision of the Council which could affect the defence or military security of the country.

Art. 9 Recourse from the decisions of the National Petroleum Council will be to the President of the Republic, within time limits to be fixed.

Art. 10. It is the duty of the National Petroleum Council:

a) to authorize, regulate and control the import, export, transport, including the construction of oil pipelines, the distribution and the commerce of petroleum and its derivatives in the national territory;

b) to authorize the installation of any refineries or depositories, deciding upon their location, the production capacity of the refineries, and the nature and quality of the refined products;

c) to establish, whenever it judges it convenient, in the defence of national economic interests and providing the refining industry with sufficient guarantees to ensure its success, the limits, maximum and minimum, of the selling prices of refined products—imported in final condition or finished in the country —having in view, as much as possible, their uniformity in the entire territory of the Republic;

d) to express opinions on the desirability of conferring authorizations for exploration and concessions for exploitation of fields of petroleum, natural gases, bituminous and pyrobituminous rocks requested by the Federal Government;

e) to express opinions on the constitution of reserves of oil-bearing zones and areas as provided in Art. 116 of Decree-Law No.366, April 11, 1939, and its sole paragraph;

f) to authorize and supervise the financial operations of the companies constituted, or which may be constituted, for the development of the industry for the refining of petroleum, imported or of national origin, whatever may be its source of extraction, in the latter case;

g) to supervise the operations of the said companies, proceeding, whenever it may deem necessary, to an examination of their accounts, for the purpose of collecting evidence which will permit the exact determination of the production cost of derivatives;

h) to organize the general norms of accounting to be adopted by the companies engaging in the refining industry, with the end to facilitate the examinations treated of above;

i) to organize and maintain a statistical service, as complete as possible, of all the operations relative to the national supply of petroleum, including the selling prices of crude petroleum and its derivatives in the national territory;

j) to suggest to the Government means which it may judge necessary for the intensification of petroleum exploration in the country and in the lowering of the cost of liquid hydrocarbons in general, of national production or imported;

k) to propose means to the Government to stimulate in the country the distillation of bituminous and pyrobituminous rocks and of solid fossil combustibles;

l) to determine among the subproducts of the distillation of petroleum those which, in accordance with the present law, ought to be included in the national supply of petroleum;

m) to verify periodically the consumption of solid or liquid hydrocarbons in the various zones of the country, the existing stocks, and fix to those concerned the quotas which they may import, within determined time limits, as well as the distribution of these quotas through the different ports of entry in the country;

n) to establish the minimum stocks of liquid hydrocarbons to be maintained permanently by the refiners or importers, in the parts of the country which may be determined, with an indication of the nature and quality of the respective products;

o) to propose the alteration of imposts and taxes of any type which concern the industry and commerce of petroleum and its subproducts, or the creation of new imposts and taxes.

Art. 11. No alteration of any of the taxes or imposts which concern the industry and commerce of petroleum and its subproducts will be made, nor new fiscal burdens be created, without prior consultation with the National Petroleum Council.

Art. 12 No international obligation which affects the commerce or the industry of petroleum and its subproducts will be assumed by the Government without prior consultation with the National Petroleum Council.

Art. 13. The National Petroleum Council will carry out, through its technical organ, which will be created, official exploration of oil and natural gas fields, as well as, when it may be judged convenient, proceeding to the development and industrialization of the respective products.

Sole Paragraph. To this effect, at an opportune moment there will be transferred to the National Petroleum Council the technical and administrative personnel and the material equipment, now in existence, destined for this work, as well as the respective budgetary credits. Funds will be assigned annually in the budget of expenses for the defrayal and development of these services.

Art. 14 The National Petroleum Council is authorized to use all means it may judge necessary to ensure the faithful obedience to the dispositions contained in the laws and regulations relative to the matter, being able to proceed to the seizure of goods and to the closing of establishments and installations of any type which it finds contravening the said laws and regulations, as well as to impose fines to a maximum of 500,000 milreis per infraction, without threat of penal action which might apply to the case.

Art. 15. To meet the expenses from the execution of the present Decree-Law, a tax of 3 milreis is created, per ton of crude oil, gasoline, kerosene, combustible oils and mineral lubricants and of any other subproduct of petroleum, at the judgment of the National Petroleum Council, imported or produced in the country with foreign or national raw material.

Sec. 1. The tax referred to in this Article will be collected, in the case of imported goods, in the respective Customs Offices and in the case of that produced in the country, by means of a way bill to the Federal Receivers, Income Boards or Tax-Collection Offices, there being, in both cases, special accounts.

Sec. 2. Crude oil, imported or of national production, in this case, whatever its source when used in national refineries as raw material, shall be exempt from the tax created in this article.

Art. 16. Expenditures with the National Petroleum Council will be charged against the credits destined for it in the budget addition of ordinary expenses and in other credit laws, the same Council being responsible for forwarding to the President of the Republic annually the budget of funds necessary for its functioning.

Sole Paragraph. The Government will open the necessary credit to cover the expenses from the founding and costs of the Council in the present fiscal year.

Art. 17. This Decree-Law will enter into effect on the date of its publication, all dispositions to the contrary being revoked.

APPENDIX II

Draft Proposal of the Petroleum Statute (1947) (Excerpts)

Title I

Of the Basic Conditions for the Development of Reserves of Fluid Hydrocarbons and Rare Gases

CHAPTER I

Of the System of Development and Its Aims

Art. 1. Reserves of petroleum and of other fluid hydrocarbons and rare gases existing in the national territory are maintained in the domain of the Union, as inalienable and imprescriptible property.

Sec. 1. Their economic development is subordinated to:

I — the resolution to guarantee to the Nation the greatest participation possible in the benefits resulting from the utilization of its petroleum wealth;

II — the requirements of the internal consumption of petroleum and derivatives;

III — the necessity of the maintenance of national reserves of crude oil;

IV — the desirability of proscribing procedures not acknowledged by the best exploitation technique.

Sec. 2. To attain these objectives the exploration and development of the reserves indicated in this article, as well as the refining of petroleum, national or imported, and the transport, by means of conduits or tankers of fluid hydrocarbons of any origin, are declared a public utility and, by Article 146 of the Constitution, of the exclusive competence of the Union.

Sec. 3. The Union will execute, by direct or contracted administration or by means of authorizations and concessions, the operations referred to in the above paragraph and will regulate the commerce of petroleum and derivatives.

Art. 2. The proprietors of the soil, deprived by this law of the exercise of the preferential rights referred to in Sec. 1 of Article 153 of the Constitution, shall be reimbursed, at their option:

a) either by the payment, upon commencement of exploitation, of an amount equivalent to the mean value, taken over the previous five years, of the areas which correspond to the respective fields of exploitation;

b) or by the annual payment of an assessment corresponding to 1 per cent of the production value, at the well mouth.

Sole Paragraph. The payments will be made for the account of the Union or of the holders of authorizations or concessions, but will always be made by the Federal Government.

Art. 3. It shall be the duty of the National Petroleum Council (CNP) to execute and to have executed this law and its regulations, always having in view the dispositions of Sec. 1 of Article 1.

. . .

CHAPTER II

Of Authorizations and Concessions

Art. 4. The following shall be the object of an authorization:

a) the reconnaissance of regions in sedimentary provinces for the selection of zones of exploration;

b) the exploration, in predetermined zones, for the selection of areas of exploitation;

c) the exploitation over a short period of small fields;

d) the export of petroleum and derivatives.

Sec. 1. In the case of line a, as many authorizations as may be required shall be granted to persons considered physically or juridically competent.

. . .

Art. 5. The following shall be the object of a concession:
a) the exploitation over a long period of large fields;
b) the transport of fluid hydrocarbons by means of conduits or tankers, by internal or coastwise shipping lines;
c) the refining of national or imported petroleum.

Art. 6. The operations referred to in the previous article may be the object of isolated or simultaneous concessions . . .

Sec. 1. The concessions discussed in lines b and c of Article 5, when they refer to products destined for national consumption, except in cases covered by the following disposition, may be held only by juridical persons of internal public right, mixed-economy companies, collective or joint-venture societies all of whose shareholders are Brazilians, or societies whose voting shares, in the minimum proportion of 60 per cent, are held by Brazilians.

Sec. 2. Once the necessities of internal consumption of gasoline are satisfied, the concessions for the refining and transport of hydrocarbons, destined for export, may be granted without the restrictions of the previous paragraph.

. . .

Art. 13. The value of the concession will include, in addition to the capital not amortized, invested in its fixed assets and installations, only the value of the fluid hydrocarbons in the fields measured and calculated by a formula adopted by the CNP; the value of reserves merely indicated or inferred and of rights and exemptions not exercised shall not be computed.

Art. 14. The duration of a concession . . . may never exceed forty years . . .

Title II
Of the Holders of Authorizations and Concessions

SECTION I

Of Natural and Juridical Persons

Art. 18. The authorizations and concessions referred to in Sec. 3 of Article 1 may be granted solely:

a) to Brazilian citizens who are in full enjoyment of their rights;

b) to societies organized in the country for such purpose, in conformity with its laws . . .

c) to juridical persons of internal federal, state or municipal law.

Sec. 1. In addition to the requisites of legal capacity, petitioners for authorizations and concessions must fulfill the requirements of technical and financial ability which shall be established by this law and which shall be at the free appraisal of the CNP.

Sec. 2. No member of the Government or public servant whose position is dependent upon or may be dependent upon the dispositions of this law may hold an authorization or concession nor participate in a society which holds them, on penalty of revocation or forfeiture.

Sec. 3. Authorizations or concessions held by societies which become subsidiaries of foreign governments and those which infringe the disposition of Sec. 1 of Article 6 shall be revoked or forefeited.

. . .

Title III
Of Exploration and Exploitation

CHAPTER I

Of the Sedimentary Provinces

Art. 24. For the purpose of this law, the non-metamorphic sedimentary basins of the national territory shall be divided into 13 provinces, delineated thus:

I—North of the Solimões, between the Colombian frontier and the Negro River.

II—South of the Solimões, between the Peruvian frontier and the divisor Juruá-Purus and the Tefé River.

III—The hydrographic basin of the Purus and Madeira, including the southern strip of the Solimões, from the divisor Juruá-Purus and the Tefé River to the Canumã River.

IV—The hydrographic basin of the Lower Amazon and the coast of Pará.

V—Maranhão, North Goiás and Southeast Pará.

VI—Piaui and Northeast Bahia.

VII—The Northeast coast, from Ceará to Alagoas.

VIII—The East coast, from Sergipe to Rio de Janeiro.

IX—The north of the hydrographic basin of the Paraná, to the north of the Paranapanema and to the east of the Paraná-Paranaíba.

X—Paraná-Santa Catarina.

XI—Rio Grande do Sul.

XII—The Northwest of the hydrographic basin of the Paraná and the hydrographic basins of the Upper Araguai and Paraguay.

XIII—The Centre-West, comprising the hydrographic basins of Upper Xingú, Upper Tapajós and Upper Madeira.

. . .

Sec. 5. When it shall be necessary, Province XIV shall be constituted, to embrace the exploration zones of the continental shelf within the national territory.

. . .

CHAPTER III

Of Exploration

. . .

Art. 29. Authorizations for exploration [are] always granted by decree . . .

. . .

Art. 35. Within six months, from the date of termination of the time limit for exploration, the holder of an authorization shall present to the CNP a detailed report, containing data which will permit of a sound judgment of the petroleum reserves of the area explored, the characteristics of the fluid hydrocarbons encountered and the possibilities for exploitation of the fields . . .

. . .

Art. 36. Within 60 days, from the presentation of the report to the CNP, the latter will pronounce on the same, publishing its decision.

. . .

CHAPTER IV
Of Exploitation

. . .

SECTION II

Of the Concession for Exploitation

. . .

Art. 49. The duration of an exploration concession, counted from the date of the transcription of the respective decree, shall be thirty years, and may be extended for up to ten years more . . .

. . .

Title IV
Of the Refining, Transport, Export and Internal Sale of Fluid Hydrocarbons

CHAPTER I

Of the Concessions for Refining and Transport

Art. 60. The Union may construct and maintain in operation works for the refining of petroleum and the treatment of natural gases, as well as conduit lines and tanker fleets for the transport of petroleum and its derivatives, by direct or contracted administration, or through concessions, under the form prescribed by this law.

. . .

CHAPTER II

Of Authorizations for Export and Internal Sale

Art. 68. The export of petroleum may be effected solely by companies authorized for the purpose by decree of the President of the Republic.

Art. 69. The CNP will fix the export quotas of each company in such a way as to assure:

a) the supply of crude oil to the refining plants whose products are destined for the supply of the internal market;

b) the refining within the country of petroleum destined for export, in progressively increasing percentages, from 25 per cent, established every five years, after a period of ten years from the first shipment.

Art. 70. Once the exigencies of internal consumption are satisfied, and once a reserve the equivalent of three times the annual consumption is assured, export shall be free, the dispositions of Sec. 1 of Article 1 and line b of Article 69 being observed.

Art. 71. The importing and internal sale of petroleum and derivatives shall be regulated by the CNP, bearing in mind the success of the national refining industry and, as much as possible, the uniformity of prices in all the territory of the Republic.

. . .

Title V
Of Fiscal Dispositions

Art. 72. Contributions . . . of contractual origin, fines incurred by holders of authorizations and concessions, profits from the direct industrialization of petroleum by the CNP and the dividends from shares held by the Union in mixed-economy companies under this law will be collected in the Bank of Brazil and placed in a special account with interest, under the name of "National Petroleum Fund".

Sec. 1. The National Petroleum Fund is destined to finance:

a) the exploration operations which the CNP performs directly, or by contract, in the zones of national reservation;

b) the subscription of shares in mixed-economy companies which may be organized, within the country, for the execution of operations [under authorization or concession];

c) the payment of indemnities of redemption or others which result from the application of this law;

d) the construction of pipelines or refineries for the interest of national security.

. . .

Art. 75. The States and Municipalities are prohibited from taxing directly or indirectly the production, commerce, distribution, transport, consumption and export of fluid hydrocarbons of whatever origin or nature.

. . .

Title VII
General and Transitory Dispositions

. . .

Art. 90. The reorganization of the CNP will proceed by special law, to the end that this organ may attend progressively to the tasks attributed to it.

. . .

Art. 95. The importers of gasoline, kerosene, combustible and lubricating mineral oils, destined for distribution and commerce, are obliged to maintain permanently, in their depots in the country, a minimum stock corresponding to 15 per cent of the quantities imported in the previous year.

Art. 96. The distributors are obliged to maintain, permanently, in their depots, in each distribution zone, a minimum stock of each product, corresponding to 15 per cent of the quantities sold in the previous year.

Art. 97. The refineries are obliged to constitute from the outset of their functioning and to maintain permanently, in the country, a minimum stock of raw material which they use corresponding to 25 per cent of their annual refining capacity.

. . .

Art. 105. Dispositions to the contrary are revoked.

APPENDIX III
Law No. 2004, October 3, 1953

The President of the Republic:
I hereby make it known that the National Congress decrees and I sanction the following law:

Chapter I
Preamble

Art. 1. The following constitute a monopoly of the Union:

I — The prospecting and development of petroleum deposits and other fluid hydrocarbons and rare gases existing in national territory;

II — The refining of national or foreign petroleum;

III — The maritime transportation of national crude oil or petroleum derivatives produced in the country, as well as the transportation by pipeline of crude oil and its derivatives and also that of rare gases whatever their origin.

Art. 2. The Union shall exercise the monopoly established in the preceding article:

I — Through the National Petroleum Council as the organ of policy and supervision;

II — Through the stock company Petróleo Brasileiro S.A. and its subsidiaries constituted as described in the present law as the executive agency.

Chapter II
Of the National Petroleum Council

Art. 3. The National Petroleum Council, an autonomous agency, directly subordinate to the President of the Republic, shall supervise the measures concerning national supplies of petroleum.

Sec. 1. By national supplies of petroleum is meant the production, importation, exportation, refining, transportation, distribution and trade in crude oil, either from wells or shale, as well as its derivatives.

Sec. 2. The utilization of other fluid hydrocarbons and rare gases is also included under the supervision of the National Petroleum Council.

Art. 4. The National Petroleum Council shall continue to be governed in its organization and functioning by existing laws and the changes arising from the present law.

Sole Paragraph. The President of the Republic shall issue new statutes for the National Petroleum Council, bearing in mind the stipulations of this article.

Chapter III
Of the Stock Company Petróleo Brasileiro S.A. and Its Subsidiaries

SECTION I

Of the Constitution of Petrobrás

Art. 5. Under the terms of this law the Union is authorized to incorporate a stock company to be denominated Petróleo Brasileiro S.A. which will use the abbreviation Petrobrás.

Art. 6. Petróleo Brasileiro S.A. shall have as its purpose the prospecting for, extracting and refining of, trade in and transportation of petroleum, either from wells or shale, and its derivatives, as well as any other correlated activities or purposes.

Sole Paragraph. The prospecting and development conducted by the company shall be in accordance with the plan organized by it and approved by the National Petroleum Council, without the formalities, area limitation exigencies and others judged to be dispensable in view of Decree-Law 3236, May 7, 1941, the Council authorizing them in the name of the Union.

Art. 7. The President of the Republic shall designate by decree the Union's representative in the acts of constitution of the company.

Sec. 1. The acts of constitution shall be preceded by:

I — The study and approval of the proposed organization of the basic services of the company, either internal or external;

II — The listing, with complete specifications, of the assets and rights to be allocated by the Union to complete its capital;

III — The drawing-up of the statutes and the prior publication of these for general knowledge.

Sec. 2. The acts of constitution shall consist of:

I — Approval of the valuation of the assets and rights listed for paying up of the Union's capital;

II — Approval of the statutes;

III — Approval of the plan of transfer of the services to pass from the National Petroleum Council to the company and the respective allocations.

Sec. 3. The company shall be constituted in a public meeting of the National Petroleum Council, the minutes of which shall incorporate the approved statutes as well as the listing and resumé of the acts of constitution, especially the valuation of assets and rights converted into capital.

Sec. 4. The constitution of the company shall be approved by Executive Decree and its contract registered by means of an authenticated copy with the Registrar of Commerce.

SECTION II

Of the Capital of Petrobrás

Art. 8. The statutes of the company shall, insofar as it is applicable, observe the norms covering stock companies. Any change in the statutes implying modification of this law shall depend on legislative authorization and, in other cases, be subject to the approval of the President of the Republic, to be made by decree.

Art. 9. The company shall have an initial capital of Cr$4,000,000,000.00 (four billion cruzeiros), divided into 20,000,000 shares of common stock, registered, of Cr$200.00 (two hundred cruzeiros) each.

Sec. 1. Up to 1957 the capital shall be increased to a minimum of Cr$10,000,000,000.00 (ten billion cruzeiros), under the terms of Article 12.

Sec. 2. The company's stock shall be common voting and preferential non-voting, inconvertible into voting stock, while any increase in capital may be either wholly or partially divided into preferential stock the issuing of which need not be governed by the restriction of Article 9, Decree-Law No. 2627, September 26, 1940.

Sec. 3. The preferential stock shall have priority in reimbursement of capital and the distribution of a minimum dividend of 5 per cent.

Sec. 4. The company's stock may be grouped in multiple blocks of 100 to 100,000 shares, the grouping and splitting of which may be regulated by the statutes according to the wishes of the stockholder.

Art. 10. The Union shall underwrite the total initial capital of the company, which shall correspond to voting stock and for the paying up of which it shall have the assets and rights it holds, relating to petroleum, including permission to use oil wells, bituminous and pyrobituminous shale deposits and natural gases; in any capital increase, it shall also underwrite common stock which shall assure it of at least 51 per cent of the voting capital.

Sec. 1. Should the value of the assets and rights referred to in this article, calculated according to the appraisal approved by the National Petroleum Council, not be sufficient to complete payment of the capital, the Union shall do so in monies.

Sec. 2. Under the hypothesis foreseen in the preceding paragraph, the National Treasury is authorized to advance, for the account of the revenue in taxes and contributions for completing the company's capital, or to carry out credit operations in anticipation of revenue, up to the amount of Cr$1,500,000,000.00 (one billion five hundred million cruzeiros).

Sec. 3. Without onus, the Union shall transfer to States and Municípios in whose territory oil wells and bituminous or pyrobituminous shales and natural gas deposits are discovered, respectively, 8 and 2 per cent of the stock referring to the value attributed to such deposits, which shall be incorporated with the capital of Petrobrás, when constituted, or later.

Art. 11. The transfer by the Union of capital stock, or the underwriting of capital increases by entities or persons to whom this right is conceded by this law, cannot under any circumstances lead to the reduction to less than 51 per cent, not only with regard to the voting stock owned by the Union but also its participation in the formation of the capital.

Sole Paragraph. Any transfer or underwriting of stock infringing this article shall be null and void, the request for such action being permitted to third parties by popular request.

Art. 12. Periodic increases of the company's capital shall be made from the revenue mentioned in the following articles.

Art. 13. That part of the revenue from the sole tax on liquid fuels referred to in Article 3 of Law No. 1749, November 28, 1952, shall be divided as follows:

I — The 40 per cent pertaining to the Union for stock in the company until the paying up of the capital called for in Sec. 1 of Article 9 is guaranteed and eventually in subscribing to debentures;

II — The 60 per cent pertaining to States, the Federal District and Municípios shall be invested:

a) in stock of the company until the paying in of capital according to plans approved by the National Petroleum Council is guaranteed, the participation of each entity to be at least proportional to the respective quota of the sole tax;

b) in debentures of the company, or stock and debentures of subsidiaries, the proportional participation of the respective contributions being always guaranteed to the States, Federal District and Municípios, the preference established in Article 40 being observed.

Sole Paragraph. The quota of the National Highway Fund, pertaining to the entities mentioned in Item II, may be withheld if any obstacle is placed in the way of the investment of the percentage specified in the referred item for the purposes and under the terms established in this article.

Art. 14. The revenue from the Import and Excise taxes on automotive vehicles and the tax on remittances abroad referring to the importation of such vehicles and their parts and accessories, shall be utilized by the Union for the underwriting of the stock and debentures of the company.

Art. 15. The proprietors of terrestrial, aquatic or air vehicles shall, up to the fiscal year 1957, contribute yearly the assessments described in the accompanying schedule and shall receive, the stipulations of Article 18 being observed, certificates which will be substituted for the company's preferred stock or debentures which will contain an express declaration of this right, the complete responsibility of the Union being guaranteed under any hypothesis for the nominal value of such bonds.

Sole Paragraph. The acts referring to automotive vehicles within the competency of the Union can only be carried out after payment of the contribution referred to in this article: the Government shall promote covenants or agreements with other public authorities so that, at the time of the yearly licensing of such vehicles within the limits of their competency, their collaboration in this respect is assured.

Art. 16. The revenue referred to in Articles 13, 14 and 15 shall be paid into the special account or accounts in the Bank of Brazil.

Sec. 1. Through the intermediary of the representative designated under the terms of Article 7, the Union may utilize the resources allocated by this law to Petrobrás, in accordance with instructions from the Ministry of Finance, to meet the respective expenditure prior to its constitution.

Sec. 2. Even though the shares corresponding to the increase of capital may not yet have been distributed, the company may utilize the special accounts referred to in this article.

Art. 17. The company may issue, up to the limit of double its paid up capital, bearer debentures, with or without the Treasury's guarantee.

SECTION III

Of the Shareholders of Petrobrás

Art. 18. Guaranteeing the preference to public national entities (government agencies), the statutes of the company may admit as stockholders only the following:

I — Public national entities;

II — The Bank of Brazil and mixed-economy companies created by the Union, States or Municípios, which companies, by force of law, are under the permanent control of the Public Authorities;

III — Native-born Brazilians, or those naturalized for more than five years and resident in Brazil, either single or married to Brazilians or foreigners when not under the regime of communal property or any other permitting the transfer of rights arising from the marriage, the purchase of common stock being limited to 20,000 shares;

IV — Private companies organized in accordance with the stipulations Article 9, line b of Decree No. 4071, May 12, 1939, the purchase of common stock being limited to 100,000 shares;

V — Private companies, Brazilian, in which only those persons indicated in Item III participate, the purchase of common stock being limited to 20,000 shares.

SECTION IV

*Of the Directorate and the
Audit Council of Petrobrás*

Art. 19. The company shall be managed by an Administrative Council with deliberative functions, and an Executive Board.

Sec. 1. The Administrative Council shall be constituted as follows:

a) one president nominated by the President of the Republic, dismissable *"ad nutum"*, having the right of veto over the decisions of the Council itself and the Executive Board;

b) three directors nominated by the President of the Republic, with a three-year mandate;

c) councillors elected by public entities, except the Union, to a maximum of three, with a three-year mandate;

d) councillors elected by private persons and companies, to a maximum of two, with a three-year mandate.

Sec. 2. The number of councillors shall be fixed in the proportion of one to every 7.5 per cent part of the company's voting capital underwritten by those mentioned in letters c and d of Sec. 1.

Sec. 3. The Executive Board shall be composed of a president and three directors nominated by the President of the Republic.

Sec. 4. Only native-born Brazilians may be members of the Administrative Council and the Audit Council.

Sec. 5. An "ex-officio" appeal to the President of the Republic, after a hearing by the National Petroleum Council, may be made against the veto of the president referred to in letter a of Sec. 1.

Sec. 6. The first three directors shall be nominated to mandates of respectively one, two and three years so that the mandate of a director expires yearly.

Art. 20. The Audit Council shall be composed of five members with three-year mandates.

Sole Paragraph. The union shall elect a representative, private persons and companies another, and the remaining public entities three, guaranteeing in this manner to each group of stockholders representing one-third of the votes the right to elect separately a member.

Art. 21. The Audit Council of Petrobrás shall have the attributes laid down in Article 127 of Decree-Law No. 2627, September 26, 1940, Decree-Law No. 2928, December 31, 1940, not being applicable.

SECTION V

Of the Benefits and Obligations Attributed to Petrobrás

Art. 22. The acts for the formation of the company and the paying in of its capital, as well as the properties it owns and the purchase of assets and real estate it makes, and still further the instruments of mandate for exercising the right to vote in General Assemblies shall be exempt from taxes and assessments and any other fiscal onus coming within the competence of the Union which shall consult with other administrative entities requesting them to extend the same favours to the company in which they will participate, within the sphere of their taxing competence.

Art. 23. The company shall enjoy exemption from import duties and taxes with regard to machinery, its parts and accessories; apparatus; tools, instruments and material for the construction, installation, expansion, improvement, functioning, development, conservation and maintenance of its installation for the purpose for which it is constituted.

Sole Paragraph. All the material and merchandise referred to in this paragraph shall be cleared through customs by virtue of an edict issued by the customshouse inspectors, with the restriction only as to similarity with national products.

Art. 24. The right to expropriate under existing legislation shall be guaranteed to the company.

Art. 25. Always depending upon prior and specific approval from the National Petroleum Council, the company may only guarantee financing, either in the country or abroad, in favour of subsidiary companies providing the operation in cases where foreign capital has no real connection.

Sole Paragraph. The Executive may extend the guarantee of the National Treasury up to 25 per cent of the respective paid-up capital to the financing obtained abroad by the company or its subsidiaries when, because of the magnitude of the operation and the high national interests involved, this becomes necessary.

Art. 26. Only when dividends reach 8 per cent shall the general stockholders' meeting determine the percentages or bonuses for the account of profits for distribution to the administration of the company.

Art. 27. The company and its subsidiaries are obliged to pay to the States and Territories in which the former develop oil and shale oil deposits, or extract gas, an indemnization equal to 5 per cent of the value of the oil, shale or gas extracted.

Sec. 1. The value of the oil and shale shall be determined by the National Petroleum Council.

Sec. 2. The payment referred to in this article shall be made each quarter.

Sec. 3. The States and Territories shall distribute 20 per cent of what they receive to the Municípios, proportionally, according to the oil production of each, this payment to be made quarterly.

Sec. 4. The States, Territories and Municípios should invest the revenue determined in this article preferably in the production of electric power and the paving of highways.

Art. 28. The Union may call on the company to carry out services conducive with its objectives, for which purpose it shall appropriate special funds.

Art. 29. The rights referring to concessions and authorizations regarding mineral oil wells, refineries and pipelines which the company receives from the Union shall be inalienable, even when Petrobrás might cede its right to utilize these as an economic value to any of its subsidiaries.

Art. 30. Should expropriation not take place, Petrobrás shall indemnify landowners at a just value for the damage caused through drilling or development.

Art. 31. In accordance with the orientation of the National Petroleum Council, Petrobrás shall maintain a minimum coefficient of oil reserves at the oil fields.

Art. 32. By March 31 of each year, Petrobrás and its subsidiaries shall forward the general accounts of the company to the General Accounting Office, referring to the preceding fiscal period, which shall be forwarded by the latter to the Chamber of Deputies and the Federal Senate.

Sole Paragraph. The General Accounting Office shall limit itself to the issuing of a report on the accounts forwarded to it. After taking knowledge of these, without judging them or the report of the Office, the National Congress shall adopt, by either of its Houses, the supervisory measures it judges convenient.

Art. 33. The management of Petrobrás and of the companies subsidiary to it are obligated to submit the information requested from it by the National Congress with regard to its acts and decisions.

Art. 34. When a stockholder is a government agency examination of the company's papers and documents will be permitted for the purpose of audit of its accounts.

Art. 35. The statutes of Petrobrás shall prescribe specific norms covering the participation of its employees in the company's profits, which shall be effective until Item IV of Article 157 of the Constitution shall, in a general manner, be covered by regulations.

SECTION VI

*Dispositions Relative to the
Personnel of Petrobrás*

Art. 36. Military or civil servants of the Union, autonomous agencies and mixed-economy companies may be employed in

Petrobrás in the managerial or technical fields, in accordance with Decree-Law No. 6877, September 18, 1944, though they may not accumulate wages, bonuses or any other advantages under the penalty of being considered to have renounced their original posts.

Sole Paragraph. In the hypothesis that the National Petroleum Council should reduce its personnel, Petrobrás shall give preference [to these] for the filling of posts or jobs according to the aptitude of the released personnel.

Art. 37. The stipulation of line c of Article 2, Decree-Law No. 538, July 7, 1938, shall not be applicable to the directors, employees and stockholders of Petrobrás, whose employees and civil servants in general, including those of the National Petroleum Council, may be stockholders.

Art. 38. The company shall contribute to the preparation of the technical personnel necessary for its services, as well as qualified labour, by means of specialized courses which it will organize; it may also award grants to educational establishments in the country or scholarships for study abroad, as well as other adequate means.

SECTION VII

Of the Subsidiaries of Petrobrás

Art. 39. The company shall operate directly or through its subsidiaries, organized with the approval of the National Petroleum Council, in which it must always hold the majority of voting stock.

Sec. 1. In the composition of the remaining capital, the same criterion established for Petrobrás shall be observed, the proportion referred to in Article 13, Item II, letter b, and the preference established in Article 40 being guaranteed.

Sec. 2. The managerial posts of the companies referred to in this article shall be prerogatives of native-born Brazilians, whenever the companies' objectives be any of the activities of the petroleum industry.

Sec. 3. In the constitution of the managerial and supervisionary boards of the subsidiaries analogous criteria to those established in this law shall be adopted, with further guarantees to government agencies having relevant interests in such companies, of representation on the executive boards.

Art. 40. Preference shall be guaranteed to the State, with the participation of its Municípios, in whose territory crude oil is extracted, in the participation of up to 20 per cent of the capital of the subsidiary companies for refining or distribution.

Sole Paragraph. Whenever the State which produces petroleum or gas manifests interest in utilizing the privilege discussed in this article, Petrobrás will transfer or attribute to it the shares it proposes to take over; time limits and conditions will be fixed so that the collaboration of the State may be effected without prejudicing the constitution and functioning of the subsidiary with which it intends to participate.

Art. 41. On authorization of the President of the Republic, issued by decree and after consultation with the National Petroleum Council, Petrobrás may, without the limitations foreseen in Article 39, associate itself with entities whose purpose is petroleum development outside national territory, providing Brazil's participation or that of Brazilian entities is covered by treaty or covenant in such cases.

Art. 42. The stipulations of Articles 22, 23, 24, 33 and 36 are equally applicable to the company's subsidiaries.

Chapter IV
Final Dispositions

Art. 43. Refineries now in operation in the country are excluded from the monopoly established by the present law and the concessions for pipelines in the same situation are maintained.

Art. 44. Authorizations to install refineries in the country issued up to June 30, 1952, shall not be prejudiced, except should they not be functioning within the periods predetermined up to the present date.

Art. 45. Authorizations to expand the capacity of the refineries covered by the two preceding articles will not be given.

Art. 46. Independently of special legislative authorization Petrobrás may participate as a stockholder in any of the refining companies covered by the preceding articles for the purpose of converting them into its subsidiaries.

Sole Paragraph. In the cases covered by the present article, Petrobrás shall purchase a minimum of 51 per cent of the stock of each company.

Art. 47. Tankers privately owned and now used in the specialized transport of petroleum and its derivatives are excluded from the monopoly established by the present law.

Art. 48. Special assessments and any others for prospecting for which concessionary companies are obligated under the form of the existing law, as well as the fines levied on those holding authorizations or concessions for any activity relating to liquid hydrocarbons, shall be for the purpose of subscribing to the stock and debentures of the company or its subsidiaries by the Union.

Art. 49. The mixed-economy companies referred to in Item IV of Article 18 shall be exclusively those existing on the date of validity of this law, their partners or stockholders being released from the proof of Brazilian nationality.

Art. 50. Whenever the National Petroleum Council must deliberate on subjects of interest to the company, the company's president shall participate in the plenary sessions though without the right to vote.

Art. 51. In the regulations covering this law the Executive shall determine the relations between the company and the National Petroleum Council.

Art. 52. The balance of the budgeted allocations and additional credits for the National Petroleum Council, for the fiscal period within which Petrobrás shall begin to operate, corresponding to services, obligations, work, equipment and purchases, or any others related to activities passing to the company, shall be handed over to the company immediately it is constituted.

Sole Paragraph. These amounts shall be credited to the paying up of the Union's capital.

Art. 53. Of the revenue from the sole tax on liquid fuels lubricants covered by Law No.1749, November 28, 1952, 48 per cent shall be allocated to the States and Federal District, the distribution being made separately for the products from national raw materials and for imported products or imported oil.

I — That part of the revenue allocated for enterprises connected with the petroleum industry (Article 3, Law No. 1749, November 28, 1952) shall be invested as called for in Article 13 of this law.

II — That part of the revenue for the National Highway Fund shall be invested in accordance with the stipulations of Law No. 302, July 13, 1938, and Law No. 1749, November 28, 1952.

Sec. 1. The revenue resulting from the products from national raw materials shall, according to the stipulations of the previous items, be distributed to the States and the Federal District in the following manner:

(1) 18 per cent proportionally to areas;
(2) 36 per cent proportionally to population;
(3) 36 per cent proportionally to consumption;
(4) 10 per cent proportionally to the production of shale or crude oil or even from condensates.

Sec. 2. The revenue resulting from imported derivatives or those produced with imported crude oil shall be distributed to the States and to the Federal District in the following manner:

(1) 20 per cent proportionally to area;
(2) 40 per cent proportionally to population;
(3) 40 per cent proportionally to consumption.

Sec. 3. The proportions of consumption foreseen in the preceding paragraphs will be calculated on the basis of the quantities consumed in each federal unit and not on the tax paid.

Sec. 4. The distribution of the quota of 12 per cent of the sole tax attributed to the municipalities shall also, where applicable, be made according to the criteria of the preceding paragraphs.

Sec. 5. The new criteria established in the present article shall be effective only in 1954.

Art. 54. Yearly the National Highway Department shall invest in highway works in the Federal Territories an amount not less than the quota which would be allocated to each should they participate in the distribution foreseen in Article 53 of this law, to be based on the collection of the preceding year.

Art. 55. The precepts of labour legislation as related to Petrobrás shall be applicable to the company's employees.

Art. 56. This law shall be effective as of the date of its publication, contrary stipulations being hereby revoked.

NOTES

Preface

1 Gen. Assis Brasil, quoted in Mário Victor, *A Batalha do Petróleo Brasileiro* (Rio de Janeiro: Civilização Brasileira, 1970), p. 53, and in Gabriel Cohn, *Petróleo e Nacionalismo* (São Paulo, 1968), pp. 11–12. The author is responsible for translations from Portuguese to English.

2 Certain interesting similarities exist between the Brazilian experience and that of Italy. The Italian government, too, prodded an oil industry into being, and the final result was similar: a vertically integrated state corporation, Ente Nazionale Idrocarburi (ENI), that was to engage directly in functions of the industry as well as control other entities. But ENI has never controlled the Italian oil industry in the way Petrobrás has the Brazilian; indeed, ENI shares the industry with the large international oil companies, and monopolizes only certain regions. Moreover, the holdings of ENI extend beyond petroleum; the diversity was greatest at the company's inception, but it has continued to act like a conglomerate while Petrobrás has restricted its activities to oil. See Charles R. Dechert, *Ente Nazionale Idrocarburi: Profile of a State Corporation* (Leiden, 1963), pp. 4–7, 10.

Introduction

1 Glycon de Paiva, "Princípios sôbre a Geologia do Petróleo", *Carta Mensal,* Vol. V (January–April 1959), pp. 37–40; Walter K. Link, *Oil and Gas Journal* (1959), pp. 289–94; Gerson Fernandes, "O Problema do Petróleo no Brasil: Situação Atual e Possibilidades Futuras", *Boletim Geográfico,* Vol. XVII (November

1960), p. 1010; Earle F. Taylor, "Geology and Oil Fields of Brazil", *Bulletin of the American Association of Petroleum Geologists*, Vol. XXXIX (August 1952), pp. 1613–26.

Chapter 1

1 Walter de Campos Birnfield, "Is There Petroleum in Brazil?", *Brazilian American*, Vol. XI, No. 273 (Jan. 17, 1925), p. 5; Euzebio Paulo de Oliveira, *História da Pesquisa de Petróleo no Brasil* (Rio de Janeiro: Oficinas Gráficas do Serviço de Publicidade Agricola, 1940), p. 129; Mauricio Vaitsman, *O Petróleo no Império de na República* (Rio de Janeiro: Seção do Livros da Empresa Gráfica o Cruzeiro, 1948), pp. 139–46; Brasil, Congresso Federal, Câmara dos Deputados, Comissão de Constituição e Justiça, *Estatuto do Petróleo. Parecer do Relator, Deputado Benedicto Costa Netto* (Rio de Janeiro, 1948), p. 52.

2 Juarez Távora, *O Petróleo do Brasil* (São Paulo: Editôra Fulgor, 1947), p. 12; Câmara dos Deputados, *Estatuto do Petróleo*, p. 52.

3 Câmara dos Deputados, *Estatuto do Petróleo*, pp. 52–3; Glycon de Paiva, "Introdução", in Oliveira, *História da Pesquisa*, p. vi; Glycon de Paiva, in Brasil, Congresso Federal, Câmara dos Deputados, *Petróleo, Projetos 1516/51, 1517/51, 1595/52: Depoimentos, Pareceres e Votos* (Rio de Janeiro, 1952), p. 51.

4 Távora, *O Petróleo do Brasil*, p. 13; Mário Victor, *A Batalha do Petróleo Brasileiro* (Rio de Janeiro: Civilização Brasileira, 1970), p. 32.

5 Victor, *A Batalha*, pp. 32–33. See also Emilio de Maya, *O Brasil e o Drama do Petróleo* (Rio de Janeiro: Livraria José Olympio Editôra, 1938). Proof of the weakness of the presidency during the Old Republic is that President Campos Sales and his successors, Rodrigues Alves and Affonso Pena, could not push a mining code into law.

6 Israel Charles White, *Final Report (Relatorio Final)* (Rio de Janeiro: Imprensa Nacional, 1908), *passim*. See also Câmara dos Deputados, *Estatuto do Petróleo*, p. 53.

7 Nonato Masson, "Yes, Nós Temos Petróleo. Brasil Pra seu Govêrno", *Jornal do Brasil*, February 28, 1962; Maya, *O Brasil*, p. 110.

8 Paiva, in *História da Pesquisa*, p. vi; Victor, *A Batalha*, p. 34.

9 Paiva, in *História da Pesquisa*, pp. vi–vii; Maya, *O Brasil*, pp. 94–95.

10 Paiva, in *História da Pesquisa*, pp. vi–vii, xi; Masson, "Yes, Nós Temos Petróleo".

11 Paiva, in *História da Pesquisa*, p. viii; Câmara dos Deputados, *Estatuto do Petróleo*, p. 54.

12 Paiva, in *História da Pesquisa*, pp. viii–ix.

13 Solidônio Leite, *O Petróleo e o Dever do Brasil* (Rio de Janeiro: J. Leite & Cia., 1927), pp. 41–42. Many of the clauses were lifted from the 1917 Mexican Constitution.

14 Birnfield, "Is There Petroleum in Brazil?", p. 7; Masson, "Yes, Nós Temos Petróleo"; Alpheu Diniz Gonsalves, *O Petróleo no Brasil* (Rio de Janeiro: Livraria José Olympio Editôra, 1963), p. 124; Ildefonso Simões Lopes, "Petróleo Nacional", *Revista do Clube de Engenharia*, Vol. XXXVI (September 1937), p. 1576.

15 Paiva, in *História da Pesquisa*, p. ix; Glycon de Paiva and Irnack C. do Amaral, "Rumos Novos em Sondagens Profundas", Brasil, Ministerio da Agricultura, *Boletim da Divisão de Fomento da Producção Mineral, Departamento Nacional da Producção Mineral*, No. 36 (1939), p. 14.

16 Távora, *O Petróleo do Brasil*, pp. 11–13; S. Fróes Abreu, "Evolução da Pesquisa de Petróleo no Brasil", *Mineração et Metalurgia*, Vol. XI, No. 61 (July 1946), p. 41; Simões Lopes, "Petróleo Nacional", p. 1576.

17 Távora, *O Petróleo do Brasil*, p. 13; *New York Times*, May 29, 1921.

18 Masson, "Yes, Nós Temos Petróleo".

19 M. A. Cremer, "Petroleum Possibilities in Brazil as Summarized in Latest Survey", *Oil and Gas Journal*, Vol. XXIII (February 12, 1925), p. 84.

20 *New York Times*, August 24, 1930; Câmara dos Deputados, *Estatuto do Petróleo*, p. 56; Dircei Lino de Mattos, "O Petróleo no Brasil", in Etienne Dalemont, *O Petróleo* (São Paulo: Coleção Saber Atual, 1961), p. 157.

21 Personal interview, Irnack Carvalho do Amaral, Rio de Janeiro, July 3, 1970. It should be noted that São Paulo was one of the few states wealthy enough to allot money for oil exploration.

22 Brasil, Ministerio da Agricultura, Industria e Commercio,

Serviço Geologico e Mineralogico do Brasil, Euzebio Paulo de Oliveira, Director, *Relatorio Annual do Director, Anno 1928* (Rio de Janeiro, 1929), pp. 3–6, 32–33, 175–80.

23 Maya, *O Brasil,* pp. 97–112; United States Department of Commerce, Bureau of Foreign and Domestic Commerce, M. A. Cremer, "Petroleum in Brazil" (Washington: Government Printing Office, 1925), pp. 4–7.

24 Maya, *O Brasil,* pp. 97–112.

25 Leite, *O Petróleo,* pp. 7, 33, 40–41, 46. For comment on state oil concessions, see Moises Rabinovitch, *Pro Solução Propria do Petróleo Econômico Nacional (Localização do Oleo Fossil no Pais)* (Rio de Janeiro, 1937–38), pp. 80–81.

26 Simões Lopes, "Petróleo Nacional", p. 1577; Simões Lopes, *O Petróleo Brasileiro* (Rio de Janeiro, 1945), pp. 16–17.

27 Câmara dos Deputados, *Estatuto do Petróleo,* p. 56; Simões Lopes, "Petróleo Nacional", p. 1577.

28 Simões Lopes, "Petróleo Nacional", pp. 1578–82; Simões Lopes, *O Petróleo Brasileiro,* pp. 25–40; Brasil, Congresso Federal, Câmara dos Deputados, Departamento dos Serviços de Taquigrafia, Directoria de Documentação e Publicidade, *Petróleo,* Vol. III (Rio de Janeiro, 1956–59), pp. 3–5.

29 "Anteprojeto de Lei sôbre as Jazidas de Petróleo", Article 12, Section 4, in Simões Lopes, *O Petróleo Brasileiro,* pp. 41–50, 163–73; Câmara dos Deputados, *Petróleo,* Vol. III, pp. 19–28, 74, 76, 80–83.

30 Câmara dos Deputados, *Petróleo,* Vol. III, pp. 126–27, 152–54; Câmara dos Deputados, *Estatuto do Petróleo,* p. 56; Simões Lopes, "Petróleo Nacional", p. 1577.

Chapter 2

1 Juarez Távora, *O Petróleo do Brasil* (São Paulo: Editôra Fulgor, 1947), pp. 14–20; Joseph E. Pogue, "Oil in Brazil", *Petroleum Engineer* (August 1951), A-45; John D. Wirth, *The Politics of Brazilian Development, 1930–1954* (Stanford: Stanford University Press, 1970), Chap. 7, covers this period well.

2 *O'Shaughnessy's Oil Bulletin,* No. 47 (March 15, 1932), p. 4; *ibid.,* No. 50 (June 15, 1932), p. 5; *ibid.,* No. 52 (August 15, 1932), p. 4.

3 Mário Victor, *A Batalha do Petróleo Brasileiro* (Rio de Janeiro: Civilização Brasileira, 1970), p. 54.

4 Alpheu Diniz Gonsalves, *O Petróleo no Brasil* (Rio de Janeiro: Livraria José Olympio Editôra, 1963), p. 141 (quoting an account by Bastos's widow in *A Tarde* [Salvador], n.d.); Pedro de Moura, in Brasil, Congresso Federal, Câmara dos Deputados, *Petróleo, Projetos 1516/51, 1517/51, 1595/52: Depoimentos, Pareceres e Votos* (Rio de Janeiro, 1952), p. 93; Nonato Masson, "Yes, Nós Temos Petróleo. Brasil Pra seu Govêrno", *Jornal do Brasil*, February 28, 1962; *New York Times*, March 26, 1933; Avelino Ignacio de Oliveira, "Situação do Problema do Petróleo no Brasil em 1938", Brasil, Ministerio da Agricultura, Departamento Nacional da Producção Mineral, Serviço de Fomento da Producção Mineral, *Boletim N. 23* (Rio de Janeiro, 1938), p. 141; "O Petróleo do Lobato", *A Tarde* (Salvador), April 1, 1933; Sílvio Fróes Abreu, in Brasil, Congresso Federal, Câmara dos Deputados, Departamento dos Serviços de Taquigrafia, Directoria de Documentação e Publicidade, *Petróleo*, Vol. III (Rio de Janeiro, 1956–59), pp. 456–57; Brasil, Congresso Federal, Câmara dos Deputados, Comissão de Constituição e Justiça, *Estatuto do Petróleo. Parecer do Relator, Deputado Benedicto Costa Netto* (Rio de Janeiro, 1948), p. 58; Mário da Silva Pinto, "Os Óleos de Lobato", Brasil, Ministério da Agricultura, Departamento Nacional da Produção Mineral, Laboratório Central da Produção Mineral, *Avulso No 3* (Rio de Janeiro, 1939), pp. 1–2, 19–24.

5 Sílvio Fróes Abreu, "O Problema do Petróleo no Brasil", *Revista Brasileira de Geografia*, Vol. I, No. 2 (September–October, 1938), pp. 53–54; personal letter from Dr. Glycon de Paiva to the author, February 7, 1968.

6 Odilon Braga, *Bases para o Inquérito sôbre o Petróleo* (Rio de Janeiro: Imprensa Nacional, 1936), pp. 41–43. It appears more than possible that the CPN directors were involved in stock speculation.

7 Edgard Cavalheiro, *Monteiro Lobato: Vida e Obra* (São Paulo: Edição da Companhia Distribuidora de Livros especialmente para a Companhia Editôra Nacional, 1955), Vol. I, pp. 416–18. The book in question was Essad-bey, *L'Épopée du Pétrole* (Paris, 1934). Monteiro Lobato is revered in Brazil for his original children's stories and translations of children's classics into Por-

tuguese. He was also a strong nationalist, and had tried unsuccessfully in the late 1920s to stir government interest in a national steel industry. Cavalheiro writes that Monteiro Lobato was "not a chauvinist with regard to petroleum", but that he did not believe "the trusts" were interested in developing Brazilian oil reserves (Cavalheiro, *Monteiro Lobato,* Vol. I, pp. 411–14).

8 Braga, *Bases para o Inquérito sôbre o Petróleo,* pp. 43–44.
9 Cavalheiro, *Monterio Lobato,* Vol. I, pp. 415–16.
10 Braga, *Bases para o Inquérito sôbre o Petróleo,* pp. 43–44.
11 *Ibid.,* pp. 44–47; Cavalheiro, *Monteiro Lobato,* Vol. I, p. 430.
12 Cavalheiro, *Monteiro Lobato,* Vol. I, p. 443; Braga, *Bases para o Inquérito sôbre o Petróleo,* pp. 50–51 (from which the *Jornal do Brasil* article is taken).
13 Távora, *O Petróleo do Brasil,* pp. 14–20; Pogue, "Oil in Brazil", A-45. See Article 118 of the 1934 Constitution, and Article 3 and Article 5, Section 3, of the Mining Code.
14 J. B. Monteiro Lobato, *O Escândalo do Petróleo e Ferro* (São Paulo: Companhia Editôra Nacional, 1936), pp. 18, 22–27.
15 Dirceu Lino de Mattos, "O Petróleo no Brasil", in Etienne Dalemont, *O Petróleo* (São Paulo: Coleção Saber Atual, 1961), pp. 157–58; Câmara dos Deputados, *Estatuto do Petróleo,* p. 58; Glycon de Paiva and Irnack C. do Amaral, "Rumos Novos em Sondagens Profundas", Brasil, Ministerio da Agricultura, *Boletim da Divisão de Fomento da Producção Mineral, Departamento Nacional da Producçao Mineral,* No. 36 (1939), pp. 64–75; Pedro de Moura, in Câmara dos Deputados, *Petróleo, Projetos 1516/51, 1517/51, 1595/52,* p. 94; Sílvio Fróes Abreu, "O Recôncavo da Bahia e o Petróleo de Lobato", *Revista Brasileira de Geografia,* Vol. I, No. 2 (April–June 1939), reprinted in *Boletim Geográfico,* Vol. VI, No. 70 (January 1949), p. 1167. Oppenheim made his assessment with precious little evidence and it would later be proven wrong. That he did not order drilling to confirm his judgement was probably due to his own rashness, a lack of equipment and the prior assumption by government geologists that the area offered no possibilities for petroleum. Oppenheim had trained as a geologist in France. His tenure in Brazil was controversial (Pedro de Moura later called him a "saboteur"), the more so on hindsight because he was wrong about Lobato. His rashness

came from "immaturity": Glycon de Paiva, answering questions put by Juarez Távora, partially quoted in Távora, "A Batalha do Petróleo; Notas à Margem do Tema 'Quem Matou Vargas' ", *Manchete*, No. 823 (January 27, 1968), p. 133 (questionnaire in the possession of the author).

16 Nelson Werneck Sodré, *História Militar do Brasil* (Rio de Janeiro: Editôra Civilização Brasileira, 1965), p. 298.

17 Monteiro Lobato, *O Escândalo do Petróleo e Ferro*, pp. 82–83.

18 Victor, *A Batalha*, pp. 92–93; Fróes Abreu, in Câmara dos Deputados, *Petróleo*, Vol. III, pp. 456–57.

19 Victor, *A Batalha*, pp. 73–82.

20 Braga, *Bases para Inquérito sôbre o Petróleo*, pp. 7, 32, 47.

21 *Ibid.*, pp. 61–78, 86–89, 91–156, 209–10; Fróes Abreu, in Câmara dos Deputados, *Petróleo*, Vol. III, p. 457.

22 Sílvio Fróes Abreu, Glycon de Paiva, and Irnack Carvalho do Amaral, *Contribuições para a Geologia do Petróleo no Recôncavo, Baía* (Rio de Janeiro: Oficinas Gráficas de Serviço de Publicidade Agricola, 1936), pp. ii–iii; Sílvio Fróes Abreu, "A Vitoria da Técnica Nacional", *Mineração e Metalurgia*, Vol. III, No. 18 (March–April 1939), pp. 367–68.

23 Sílvio Fróes Abreu, in Câmara dos Deputados, *Petróleo*, Vol. III, pp. 460–61.

24 Monteiro Lobato, *O Escândalo do Petróleo e Ferro*, pp. 158–61.

25 Wirth, *Politics of Development*, p. 144.

26 Távora, *O Petróleo do Brasil*, pp. 14–20; Pogue, "Oil in Brazil", A-45. See Article 143, Section 1, and Article 119, Section 1, of the 1937 Constitution.

27 Wirth, *Politics of Development*, p. 138–39.

28 *Ibid.*, p. 145; "O Petróleo e a Defesa Nacional; Memorial do General Júlio C. Horta Barbosa ao Ministro da Guerra, em 30-1-1936", Câmara dos Deputados, *Petróleo*, Vol. II, p. 6.

29 Peter Seaborn Smith, "Bolivian Oil and Brazilian Economic Nationalism", *Journal of Inter-American Studies and World Affairs*, Vol. XIII, No. 2 (April 1971), pp. 166–69; Alvaro Lins, *Rio-Branco* (Rio de Janeiro: Livraria José Olympio Editôra, 1945), Vol. I, pp. 366–74; Vol. II, pp. 401–35; Gordon Ireland, *Boundaries, Possessions, and Conflicts in South America* (Cambridge: Harvard Uni-

versity Press, 1938), pp. 48–53; [Refinaria e Exploração de Petróleo União S.A.], *Petróleo Boliviano e Mercado Brasileiro* (São Paulo, 1956), pp. 13–14, 35–37.

30 Paiva and Amaral, "Rumos Novos em Sondagens Profundas".

31 J. Soares Pereira, "Getúlio Vargas e o Petróleo Brasileiro", in Getúlio Vargas, *A Política Nacionalista do Petróleo no Brasil* (Rio de Janeiro: Livraria José Olympio Editôra, 1964), pp. 34–35; Távora, *O Petróleo do Brasil*, p. 21; "Brazil Establishes National Petroleum Council", *World Petroleum*, Vol. IX, No. 9 (September 1938), pp. 39, 48.

32 *New York Times*, April 30, 1938.

33 Glycon de Paiva, in Câmara dos Deputados, *Petróleo, Projetos 1516/51, 1517/51, 1595/52*, p. 51.

34 Cavalheiro, *Monteiro Lobato*, Vol. I, pp. 452–53, 464, 468–69. To his credit, Monteiro Lobato never gave up. He sent an impassioned letter to Vargas in May 1940, trying to "open his eyes" about the crime of petroleum in which officialdom "was still persisting". After Vargas had tried to win Monteiro Lobato over with offers of lucrative sinecures, he jailed him as a political prisoner on March 30, 1941.

35 Wirth, *Politics of Development*, pp. 142–43.

36 Fróes Abreu, "O Problema do Petróleo no Brasil", pp. 57–62.

37 Moises Rabinovitch, *Pro Solução Propria do Petróleo Econômico Nacional (Localização do Oleo Fossil no Pais)* (Rio de Janeiro, 1937–38), p.90.

38 Diniz Gonsalves, *O Petróleo no Brasil*, p. 130; Fróes Abreu, "O Problema do Petróleo no Brasil", p. 57.

39 Diniz Gonsalves, *O Petróleo no Brasil*, p. 124; Pedro de Moura, in Câmara dos Deputados, *Petróleo, Projetos 1516/51, 1517/51, 1595/52*, pp. 94–95; Euzebio Paulo de Oliveira, *História da Pesquisa de Petróleo no Brasil* (Rio de Janeiro: Oficinas Gráficas do Serviço de Publicidade Agricola, 1940), p. 147.

40 Pedro de Moura, in Câmara dos Deputados, *Petróleo, Projetos 1516/51, 1517/51, 1595/52*, p. 94.

41 Glycon de Paiva, "Introduction", in Oliveira, *História da Pesquisa*, p. xii; *New York Times*, January 25, 1939; "Exploratory

Activities in Brazil", *Petroleum Engineer* (October 1941), p. 76.

42 Távora, *O Petróleo do Brasil*, p. 22. Poor Oscar Cordeiro! This decree barred him from a well he regarded as his own, gave him no compensation or recognition for his effort, and left him bitter for the remainder of his life. In 1965, Petrobrás (the state-monopoly oil company created in 1954) belatedly made him an honorary technician and gave him a modest salary for life. He died in July 1970. The widow of Manoel Inácio Bastos, the original discoverer (despite Cordeiro's claims to the contrary), received a salary one-half that of Cordeiro's for the rest of her life.

43 "Relatório do General J. C. Horta Barbosa ao Presidente da República sôbre sua Viagem ao Prata, em abril de 1939", Câmara dos Deputados, *Petróleo*, Vol. II, pp. 279–303. ANCAP represented Administración Nacional do Combustibles, Alcohol y Portland; it was a state-monopoly company for the refining of imported crude oil. Horta Barbosa's report probably influenced the form of the CNP, as set up in Decree-Law 538 (July 7, 1938): see Appendix I.

44 Távora, *O Petróleo do Brasil*, p. 22. The transfer was effected by Decree Laws 1217 (April 24, 1939) and 1369 (June 23, 1939).

45 *New York Times*, July 15, 1939.

46 Paiva, in *História da Pesquisa*, p. xii. Gonzago de Campos had suggested such a policy in 1919, and it was finally adopted because of the persistence of Irnack Carvalho do Amaral. In mid-1939 the CNP awarded the first exploration contract to the Drilling and Exploration Co., Inc., of Los Angeles: *World Petroleum*, Vol. X, No. 7 (July 1939), p. 45; Wallace A. Sawdon, "Brazilian Government Begins Aggressive Exploratory Program", *Petroleum Engineer* (October 1939), p. 128.

47 Wirth, *Politics of Development*, p. 151.

48 *Ibid.*, pp. 151–55; "Relatório do General J. C. Horta Barbosa", Câmara dos Deputados, *Petróleo*, Vol. II, pp. 279–303.

49 Wirth, *Politics of Development*, pp. 153–54; "Brazilian Market Upset by Tariffs", *World Petroleum* (January 1937), pp. 25–26; Leonard M. Fanning, *American Oil Operations Abroad* (New York: McGraw-Hill, 1947), p. 120; *New York Times*, November 2 and 20, 1939.

50 Francisco Prestes Maia, "A Indústria do Petróleo", *Digesto Econômico*, No. 83 (October 1951), in Câmara dos Deputados, *Petróleo*, Vol. II, p. 672; Fanning, *American Oil*, p. 120.

51 "Active Development in South America Despite Uncertain Outlook", *World Petroleum*, Vol. XI, No. 7 (July 1940), p. 45.

52 Paiva, in *História da Pesquisa*, pp. xii–xiv; "Exploratory Activities in Brazil", *Petroleum Engineer* (October 1941), p. 76.

53 "Geophysical Exploration in Brazil", *Petroleum Engineer* (October 1942), p. 122.

54 Decree-Law 1985 (January 29, 1940), in Brasil, Ministério das Minas e Energia, Conselho Nacional do Petróleo, *Legislação do Petróleo* (Rio de Janeiro: Britânia Editôra, 1964), pp. 73–107. The relevant clauses are Articles 5 and 6.

55 "Getúlio Vargas e o Petróleo Brasileiro", *Última Hora*, January 4, 1957; "O Memorandum e a Proposta da 'Standard Oil of Brazil' ", Câmara dos Deputados, *Petróleo*, Vol. II, pp. 304–6, Horta Barbosa's note is dated August 2, 1940. The contents of the memorandum were not revealed until 1957.

56 "O Memorandum e a Proposta da 'Standard Oil of Brazil' ", Câmara dos Deputados, *Petróleo*, Vol. II, pp. 304–10. See Article 143, Section 1, of the Constitution, and Article 6 of the Code of Mines.

57 "Getúlio Vargas e o Petróleo Brasileiro", *Última Hora*, January 4, 1957.

58 Decree-Law 3236 (May 7, 1941), in Conselho Nacional do Petróleo, *Legislação do Petróleo*, pp. 109–19. Presumably Vargas decreed the law in recognition that Brazil now had oil in commercial quantities (Candeias, the first commercial field, having just been discovered), and after years of pressure to enact a special law for this special commodity.

59 Paiva, in *História da Pesquisa*, pp. xii–xiv; "Exploratory Activities in Brazil", *Petroleum Engineer* (October 1941), p. 76.

60 Sílvio Fróes Abreu, "Aspectos Geográficos, Geológicos e Politicos da Questão do Petróleo no Brasil", *Revista Brasileira de Geografia*, Vol. VIII, No. 4 (October–December 1946), pp. 524–25; Sílvio Fróes Abreu, "Problemas de Combustível no Brasil", *Mineração e Metalurgia*, Vol. VI, No. 36 (January 1, 1943), p. 281; Pedro

de Moura, in Câmara dos Deputados, *Petróleo, Projetos 1516/51, 1517/51, 1595/52,* p. 95. The Lobato field was under 100,000 barrels capacity.

61 Mattos, "O Petróleo no Brasil", in Dalemont, *O Petróleo,* p. 159; Brasil, Ministério das Minas e Energia, Conselho Nacional do Petróleo, *Relatório de 1946* (Rio de Janeiro, 1948), p. 13. Strong nationalists criticized the government for its apparent apathy; the most extreme of them accused the CNP of having sold out to "the trusts" (President Vargas imprisoned Monteiro Lobato in 1941 for having made such a charge).

62 Miragaia Pitanga, "Conselho Nacional do Petróleo em 1948", in Diniz Gonsalves, *O Petróleo no Brasil,* p. 134.

63 *New York Times,* July 16 and August 14, 1941; Decree-Law 737 (September 23, 1938), in Conselho Nacional do Petróleo, *Legislação do Petróleo,* pp. 31–33, made compulsory the addition of anhydrous alcohol to gasoline of national production; in 1931, alcohol had been ordered added to imported gasoline: Georg Kaltenbrunner, "Compulsory Use of Alcohol in Motor Fuel", *World Petroleum,* Vol. IV, No. 4 (April 1933), p. 121; *O'Shaughnessy's Oil Report,* Vol. V, No. 1 (May 1930), p. 13.

64 *New York Times,* July 22 and August 15, 17, and 28, 1941.

65 "Ofício do Presidente do Conselho Nacional do Petróleo, General Julio C. Horta Barbosa, ao Presidente da República, Sr. Getúlio Vargas (18-7-41)", Câmara dos Deputados, *Petróleo,* Vol. II, pp. 311–17; Gral. J. C. Horta Barbosa, *Problemas do Petróleo no Brasil* (Rio de Janeiro, 1947), pp. 33–34. The Standard Oil proposal was made to Vargas in a letter dated May 29, 1941.

66 "Getúlio Vargas e o Petróleo Brasileiro", *Última Hora,* January 4, 1952.

67 Horta Barbosa, *Problemas do Petróleo no Brasil,* pp. 34–35; *O Globo,* May 17, 1952.

68 "Política Nacional do Petróleo. Ofício do Presidente do Conselho Nacional do Petróleo, General Julio C. Horta Barbosa, ao Ministro da Guerra, General de Divisão Eurico Gaspar Dutra, encaminhando cópia da exposição por ele feita, em 20-10-42, ao Presidente da República, sôbre opiniões externadas pelo Sr. Aluízio de Lima Campos, em artigo publicado no *Diario*

Carioca", Câmara dos Deputados, *Petróleo*, Vol. II, pp. 321–22 (Horta Barbosa cites Wallace E. Pratt, *Oil in the Earth* [Lawrence, Kansas, 1942], pp. 85–86).

69 *New York Times,* November 20, 1941, April 19 and 23–25, May 22, June 30, and July 12 and 23, 1942, October 30, 1943, and January 30, 1944.

70 Kenneth J. Langley, "Brazil Shows Good Geologic Possibilities of Commercial Quantities of Petroleum", *Oil and Gas Journal* (December 29, 1945), p. 163.

71 Fróes Abreu, "Aspectos Geográficos", p. 527.

72 *Ibid.,* pp. 524–25; Fróes Abreu, "Problemas de Combustível no Brasil", p. 281; Pedro de Moura, in Câmara dos Deputados, *Petróleo, Projetos 1516/51, 1517/51, 1595/52,* p. 95.

73 Horta Barbosa later alleged that "the trusts" had gained influence over Vargas, who had given the CNP president little alternative but to resign. He also said later that public opinion was not prepared for his policies by 1943: Mattos, "O Petróleo no Brasil'", in Dalemont, *O Petróleo,* p. 159.

Chapter 3

1 Juarez Távora, *O Petróleo do Brasil* (São Paulo: Editôra Fulgor, 1947), p. 23.

2 Caio Pandiá Guimaraens, "O Código de Minas", *Mineração e Metalurgia,* Vol. VII, No. 41 (November–December 1943), p. 295.

3 Decree-Law 6230 (January 29, 1944), cited in Távora, *O Petróleo do Brasil,* pp. 23–24.

4 Brasil, Ministério das Minas e Energia, Conselho Nacional do Petróleo, *Relatório de 1944* (Rio de Janeiro, 1945), pp. 1–3.

5 Sílvio Fróes Abreu, "Aspectos Geográficos, Geológicos e Políticos da Questão do Petróleo no Brasil", *Revista Brasileira de Geografia,* Vol. VIII, No. 4 (October–December 1946), pp. 524–25; Sílvio Fróes Abreu, "Problemas de Combustível no Brasil", *Mineração e Metalurgia,* Vol. VI, No. 36 (January 1, 1943), p. 281; Pedro de Moura, in Brasil, Congresso Federal, Câmara dos Deputados, *Petróleo, Projetos 1516/51, 1517/51, 1595/52: Depoimentos, Pareceres e Votos* (Rio de Janeiro, 1952), p. 95.

6 Miragaia Pitanga, "Conselho Nacional do Petróleo em 1948", in Alpheu Diniz Gonsalves, *O Petróleo no Brasil* (Rio de

Janeiro: Livraria José Olympio Editôra, 1963), p. 134; Conselho Nacional do Petróleo, *Relatório de 1946* (Rio de Janeiro, 1948), p. 13.

7 Conselho Nacional do Petróleo, *Relatório, 1° triênio, 1938−41* (typewritten, Rio de Janeiro, 1941), p. 95.

8 [Odilon Braga], *Ante-projeto do Estatuto do Petróleo com o Parecer do Relator* (Rio de Janeiro: Imprensa Nacional, 1948), pp. 7, 15−16; *Jornal do Comércio,* February 19, 1948. The CNP had had before it "for sometime [sic] an offer by the Standard Oil Company of New Jersey to invest Cr$200,000,000 in a search for oil over a period of ten years": "Brazil Looking to Accelerated Oil Development", *World Petroleum,* Vol. XVII, No. 7 (July 1946), p. 40. The figure would be the equivalent of $10 million.

9 "Brazil Looking to Accelerated Oil Development", pp. 40−41; John D. Wirth, *The Politics of Brazilian Development, 1930−1954* (Stanford: Stanford University Press, 1970), p. 161.

10 Brasil, Ministério das Minas e Energia, Conselho Nacional do Petróleo, *Resoluções do Plenário, 1945 a 1967,* Vol. I (mimeographed, Rio de Janeiro, n.d.), pp. 1−3; Brasil, Congresso Federal, Câmara dos Deputados, Departamento dos Serviços de Taquigrafia, Directoria de Documentação e Publicidade, *Petróleo,* Vol. III (Rio de Janeiro, 1956−59), pp. 209−10.

11 Carlos Eduardo Paes Barreto, "A Refinaria de Petróleo de Mataripe", *Revista do Clube de Engenharia,* No. 171 (November 1950), p. 337; "Brazil's Latest Discovery Spurs Interest in Oil Development", *World Petroleum,* Vol. XVII, No. 12 (November 1946), pp. 58−59.

12 Glycon de Paiva, "Suprimento de Petróleo para o Brasil", *Mineração e Metalurgia,* Vol. X, No. 57 (January−February 1946), pp. 125−26; Fróes Abreu, "Aspectos Geográficos", p. 530; "Esbôço Histórico das Pesquisas de Petróleo no Brasil pelo Prof. Sílvio Fróes de Abreu", Câmara dos Deputados, *Petróleo,* Vol. III, p. 456.

13 Thomas E. Skidmore, *Politics in Brazil, 1930−1964: An Experiment in Democracy* (New York: Oxford University Press, 1967), pp. 49−50, 56, 64−65, 69.

14 Távora, *O Petróleo do Brasil,* pp. 25−26; Brasil, Congresso Federal, Câmara dos Deputados, Comissão de Constituição e

Justiça, *Estatuto do Petróleo. Parecer do Relator, Deputado Benedicto Costa Netto* (Rio de Janeiro, 1948), p. 107; Braga, *Ante-projeto do Estatuto do Petróleo,* p. 7.

15 Wirth, *Politics of Development,* pp. 161–64; Conselho Nacional do Petróleo, *Resoluções do Plenário,* Vol. I, pp. 4–10, 17–24, 55–60.

16 Conselho Nacional do Petróleo, *Relatório 1947* (Rio de Janeiro, 1948), pp. viii–xi.

17 Câmara dos Deputados, *Estatuto do Petróleo,* p. 82.

18 "Esclarecimentos sobre o Ante-projeto do Estatuto do Petróleo", *Ante-projeto do Estatuto do Petróleo,* pp. 7–12. Pertinent excerpts of the draft bill are reproduced in Appendix II; the full text may be found in Câmara dos Deputados, *Petróleo,* Vol. III, pp. 290–319.

19 Major Ituriel do Nascimento, *O Brasil e o Seu Petróleo* (Rio de Janeiro: Livraria Clássica Brasileira, 1948, reprinted from the *Revista Militar Brasileira,* ns. 3 e 4, 2° semestre de 1947), pp. 4, 25; J. F. de Barros Pimentel, *O Problema do Petróleo no Brasil* (Rio de Janeiro: Imprensa Nacional, 1948), pp. 9, 13; *Ante-projeto do Estatuto do Petróleo,* pp. 7–8.

20 Távora, *O Petróleo do Brasil,* pp. 29–40, 42–46; the same program formed the basis of three addresses Távora gave to the Clube Militar: *Revista do Clube Militar,* No. 82 (May–June 1947), pp. 68–71; *ibid.,* No. 84 (September–October 1947), pp. 39–40, summarizes an address on the same theme Távora gave to the Clube Naval, September 15; Raimundo D. Padilha, "O Problema do Petróleo Nacional (Relatório apresentado ao diretório nacional do Partido de Representação Popular e aprovado em 30-6-48)", Câmara dos Deputados, *Petróleo,* Vol. II, pp. 603–4; *New York Times,* November 15, 1947. Távora's other addresses and thoughts on the subject are contained in Juarez Távora, *O Problema Brasileiro do Petróleo: Ensaio de Solução Objetiva* (Rio de Janeiro: Livraria José Olympio Editôra, 1948), and *Petróleo para o Brasil* (Rio de Janeiro: Livraria José Olympio Editôra, 1955). See also Geraldo Rocha, "O Conselho contra o Petróleo", *Mineração e Metalurgia,* Vol. XII, No. 69 (July–September 1947), pp. 128–29.

21 Horta Barbosa gave three major addresses on this theme, two in the Clube Militar, on July 30 and August 6, and one on October 16 in the Argentine Engineering Institute. See Horta

Barbosa, "O Problema do Petróleo no Brasil", *Revista do Clube Militar*, No. 83 (July–August 1947), pp. 93–95; Horta Barbosa, *Problemas do Petróleo no Brasil* (Rio de Janeiro, 1947); "Los Problemas del Petróleo en el Brasil", *Nuestro Petróleo debe ser Explotado por el Gobierno* (Buenos Aires, 1948).

22 Horta Barbosa, *Problemas do Petróleo no Brasil*, p. 7.

23 *Ibid.*, pp. 4–46; Horta Barbosa, "Los Problemas del Petróleo en el Brasil", pp. 32–38; Padilha, in Câmara dos Deputados, *Petróleo*, Vol. II, pp. 600–1.

24 "Projeto N° 382-1947: Cria o Instituto Nacional do Petróleo e dá Outras Providências", Câmara dos Deputados, *Petróleo*, Vol. III, pp. 171–78; Câmara dos Deputados, *Estatuto do Petróleo*, pp. 80–81. Distribution of petroleum products was then performed exclusively by five international oil companies: Standard of New Jersey, Atlantic, Texaco, Gulf, and Shell.

25 Câmara dos Deputados, *Estatuto do Petróleo*, pp. 80–82; Matos Pimenta, "A Campanha do Petróleo e os Comunistas", *Jornal de Debates*, December 17, 1948.

26 Câmara dos Deputados, *Estatuto do Petróleo*, pp. 81–82.

27 Eng. Fernando Luiz Lôbo Carneiro, "O Brasil Pode, Ele Próprio, Explorar o Seu Petróleo", *Jornal de Debates*, August 22, 1947.

28 Matos Pimenta, in *Jornal de Debates*, December 17, 1948.

29 "Petróleo, Monopólio e Imperialismo", *Jornal de Debates*, October 10, 1947.

30 Lôbo Carneiro, "A Questão do Petróleo no Brasil", *Jornal de Debates*, October 31, 1947. The address had been given October 7, and the statement cited was from the *Correio da Manhã* of October 11, 1947.

31 Wirth, *Politics of Development*, Chaps. 8 and 9, examines these groups from 1947 to 1954; see also Cohn, *Petróleo e Nacionalismo*, for a more sociological analysis of the same groups (readers should also consult Wirth's excellent criticism of Cohn, pp. 246–47).

32 Alfred Stepan, *The Military in Politics: Changing Patterns in Brazil* (Princeton: Princeton University Press, 1970), is the best analysis to date of the Brazilian military.

33 See Albert Breton, "The Economics of Nationalism", *Journal of Political Economy* (1964), pp. 376–86, for a penetrating

discussion of nationalism as a job-creating policy. Breton (pp. 378–79) makes the point that nationalism will create middle- rather than lower-class jobs, since the latter are either already held by nationals or deemed "unimportant"; this would make nationalism appealing to university students, who will enter, at the least, the middle class.

34 Artur Bernardes, Session of June 24, 1952, in Câmara dos Deputados, *Petróleo*, Vol. VII, p. 496. Bernardes emotionally warned Brazilians not to condemn future generations to starvation by alienating oil reserves.

35 Skidmore, *Politics in Brazil*, pp. 88–90.

36 Personal interview, Gen. Juarez Távora, Rio de Janeiro, July 8, 1970.

37 Skidmore, *Politics in Brazil*, pp. 88–90, 164–65.

38 Peter R. Odell, "The Oil Industry in Latin America", in Edith T. Penrose, *The Large International Firm in Developing Countries: The International Petroleum Industry* (London: Allen & Unwin, 1968), pp. 274–75.

39 Artur Bernardes, in Câmara dos Deputados, *Petróleo*, Vol. VII, p. 496. The "O petróleo é nosso" campaign must be seen in the light of current public knowledge about Brazilian petroleum geology, which by 1947 was still very limited at best. The CNP, restricted by small budgets, had not been able to explore very much; moreover, its annual reports, which were not distributed to the public, contained little geological information. The CNP did not have a public-relations department. Thus, there was little information for the informed reader, apart from the occasional article in a technical journal; for the uninformed public, it is safe to say that there was no comprehensible body of facts available. The Brazilian population, with few exceptions, simply believed that the national subsoil was impregnated with oil.

40 "Exposição do Relator", *Ante-projeto do Estatuto do Petróleo*, pp. 35–56.

41 "Ofício N° 702, de 30-12-1947, do Conselho de Segurança Nacional, ao Sr. Presidente da República", Câmara dos Deputados, *Petróleo*, Vol. III, pp. 320–21.

42 "Apreciações da Comissão de Investimentos sôbre o Estatuto do Petróleo", Câmara dos Deputados, *Petróleo*, Vol. III, pp. 468–78; "Comissão de Investimentos: Sugestões da Comissão

de Investimentos para a Elaboração do Anteprojeto do Estatuto do Petróleo", *ibid.*, Vol. IV, pp. 17–19.

43 "Comentários sôbre o Anteprojeto do Estatuto do Petróleo Preparado pela Comissão do Anteprojeto da Legislação do Petróleo—Herbert Hoover, Jr., e Arthur A. Curtice", Câmara dos Deputados, *Petróleo*, Vol. III, pp. 398–403; "Discussão Detalhada dos Dispositivos Pertinentes ao Anteprojeto de Estatuto do Petróleo Preparado pela Comissão de Anteprojeto da Legislação do Petróleo", *ibid.*, pp. 414, 455. In discussing his commission's ideas before the Engineering Club prior to the appearance of the bill, Odilon Braga had stressed that companies that explored a region would be guaranteed the right of exploitation. The bill, perhaps as a concession to popular nationalism, did not give such a guarantee.

44 Távora, "Defesa de uma Solução Objetiva para o Problema do Nosso Petróleo", *Petróleo para o Brasil*, pp. 135–36, 146 (an address to the Clube Militar, June 23, 1948). Hoover and Curtice made the same point about the predominance of the small operator in American exploration: "Discussão Detalhada", Câmara dos Deputados, *Petróleo*, Vol. III, p. 414.

45 A useful summary of contemporary press opinion may be found in "Parecer da Comissão de Indústria e Comércio da Câmara dos Deputados. Relator: Deputado Armando Fontes", Câmara dos Deputados, *Petróleo*, Vol. IV, p. 195. *O Jornal* was the leader of Assis Chateaubriand's *Diários Associados*, and espoused the opposite thesis—foreign-capital participation in the national oil industry—after 1950.

46 Jorge Abdalla Chamma, "Nacionalismo e Petróleo", *Tribuna* (Corumbá), April 25, 1948, in Chamma, *Por um Brasil Melhor* (Rio de Janeiro: Livraria Clássica Brasileira, 1955), pp. 118–20.

47 *Folha da Manhã* (São Paulo), November 12, 1947.

48 Juracy Magalhães, "Petróleo, Fonte de Libertação ou de Escravidão dos Povos", *O Jornal*, cited in "Transcrição nos Anais da Câmara dos Deputados de Artigos do Deputado Juracy Magalhães, Publicados na Imprensa Carioca, Sessão de 16-6-1948", Câmara dos Deputados, *Petróleo*, Vol. II, p. 559. Magalhães was then an army colonel.

49 Hélio de Lacerda, interviews in the São Paulo papers *A*

Hora, April 20 and August 4, 1948, and *O Dia,* June 30, 1948, in Lacerda, *Petróleo e Outros Problemas. Democracia, Reforma Agrária, Industrialização. Artigos, Discursos, Conferências e Entrevistas, 1947–1948* (São Paulo, n.d.), pp. 23–24, 27, 35.

50 "O Problema do Petróleo no Brasil: Contra as Ofensivas Imperialistas dos Trusts", *Revista do Clube Militar,* No. 87 (February–March 1948), pp. 19-21. Carvalho wrote *Petróleo! Salvação ou Desgraça do Brasil?,* published by the Petroleum Centre. Captain Humberto Freire de Andrade edited the *Revista* in the 1948–50 period, under the Cultural Department headed by Colonel Henrique Cunha: Nelson Werneck Sodré, *Memórias de um Soldado* (Rio de Janeiro: Civilização Brasileira, 1967), p. 299.

51 Capitão X, "O Problema do Petróleo no Brasil: Comentários à Margem do Ante-projeto do Estatuto do Petróleo", *Revista do Clube Militar,* No. 88 (April 1948), pp. 41–45; *ibid.,* No. 89 (May–June 1948), pp. 58–62. The Review often portrayed Brazilian oil men as "The Soldiers of Petroleum" or, to quote one example, "the new soldiers of Independence, of the Economic Independence of Brazil", *ibid.,* No. 89 (May–June 1948), p. 153.

52 "Explorações Políticas em torno do Petróleo", *Mineração e Metalurgia,* Vol. XII, No. 72 (March–April 1948), pp. 263–64; "Técnica e Capital Estrangeiros", *ibid.,* Vol. XV, No. 86 (July–August 1950), p. 37.

53 R. Descartes de Garcia Paula, "Petróleo e Minerais Estratégicos em face da Economia Brasileira", *Revista do Clube de Engenharia,* No. 170 (October 1950), pp. 309–16; *ibid.,* No. 171 (November 1950), pp. 349–52.

54 "Discurso do Deputado Hermes Lima", Câmara dos Deputados, *Petróleo,* Vol. IV, pp. 301–27. Lima led a left-wing faction of the UDN (with Domingos Vellasco), known as the "Esquerda Democrática", which had by this time broken away; Lima had joined the PSD: Skidmore, *Politics in Brazil,* p. 353.

55 Padilha, in Câmara dos Deputados, *Petróleo,* Vol. II, pp. 611–13; Skidmore, *Politics in Brazil,* p. 124.

56 "Discurso do Deputado Euzébio Rocha", Câmara dos Deputados, *Petróleo,* Vol. IV, pp. 373–405. This speech was given on June 3 and 4, 1948.

57 "Discurso do Deputado Artur Bernardes", Câmara dos

Deputados, *Petróleo*, Vol. IV, pp. 414–28. The speech was given on October 5, 1948. He had helped to block and finally to cancel Percival Farquhar's Itabira iron-ore concession in the 1920s and 1930s: Skidmore, *Politics in Brazil*, p. 109; Wirth, *Politics of Development*, Chap. 4; Werner Baer, *The Development of the Brazilian Steel Industry* (Nashville: Vanderbilt University Press, 1969), Chap. 4.

58 "Discurso do Deputado Pereira da Silva", Câmara dos Deputados, *Petróleo*, Vol. IV, pp. 406–13.

59 Padilha, in Câmara dos Deputados, *Petróleo*, Vol. II, pp. 620–23, 628.

60 Othon Henry Leonardos, "Por que so Petróleo Estatal?", *Mineração e Metalurgia*, Vol. XV, No. 86 (July–August 1950), p. 47.

61 Padilha, in Câmara dos Deputados, *Petróleo*, Vol. II, p. 628.

62 Pedro de Moura and Gerson Fernandes, "Petroleum Geology in the State of Bahia", *International Geological Congress, 19th, Algeria*, C. R. Sec. 14, F.16 (1953), pp. 65–84; W. B. Sherman, "Petroleum Possibilities of the Gondwana System of Brazil", *ibid.*, pp. 85–96; D. S. Oddone, "Oil Prospects in the Amazon Region", *ibid.*, pp. 247–72. These comprised the first official data on Brazil's oil geology released outside the country since the creation of the CNP.

63 "O Petróleo no Plano SALTE", Câmara dos Deputados, *Petróleo*, Vol. IV, pp. 276–84; Alfredo Marques Vianna, "Apresentação", in Getúlio Vargas, *A Política Nacionalista do Petróleo no Brasil* (Rio de Janeiro: Livraria José Olympio Editôra, 1964), pp. 17–18; "Refinaria do Petróleo Artur Bernardes", *Revista do Clube de Engenharia*, No. 225 (May 1955), p. 28.

The SALTE Plan was the result of years of debate with the United States over the best development plan for Brazil. During the Second World War the "Cooke Mission" had examined Brazilian resources and recommended steps for petroleum similar to those now taken in SALTE. Similarly, the postwar "Abbink Mission" had urged development of national oil reserves: Skidmore, *Politics in Brazil*, pp. 45, 72; William J. Kemnitzer, "Petróleo", in Fundação Getúlio Vargas, *A Missão Cooke no Brasil* (Rio de Janeiro, 1949), pp. 164, 169, 172; Octávio Gouvêa de Bulhões, *À Margem de um Relatório: Texto das Conclusões da Comis-*

são Mista Brasileira—Americana de Estudos Econômicos (Missão Abbink) (Rio de Janeiro: Edições Financeiras, 1950), pp. 230–32; New York Times, March 6, 1949.

64 Wirth, Politics of Development, p. 179; Skidmore, Politics in Brazil, p. 71. For a liberal critique of the SALTE Plan, see Joseph Pogue, Petroleum Engineer (August 1951), A-45-48. Pogue was a petroleum consultant with the Chase National Bank, which published this same assessment in 1951.

65 New York Times, January 20 and July 17 and 30, 1949. The decision to build at São Paulo rather than at Belém was prompted by contemporary Brazilian beliefs about the economics of industrialization: that it made more sense to concentrate manufacturing in the areas of greatest consumption, thereby reducing shipping costs, than to spread production throughout the country.

66 World Petroleum, Vol. XXI, No. 1 (January 1950), p. 44; R. G. Walker, "Oil Industry Progress in Brazil", ibid., No. 2 (February 1950), p. 39. The original applications had been made in 1946: see pp. 42, 44–5.

67 New York Times, July 5, 1949.

68 World Petroleum, Vol. XXI, No. 2 (February 1950), p. 46.

69 Skidmore, Politics in Brazil, p. 71; Werner Baer, Industrialization and Economic Development in Brazil (New Haven: Yale University Press, 1965), pp. 61–63.

70 World Oil (July 15, 1950), p. 120; ibid. (July 15, 1951), p. 178. Up to 1950, the CNP used all Brazil's oil production to provide fuel for drilling rigs and equipment. By the end of 1950, the CNP's 5,000 b.p.d. Mataripe refinery had begun operations (it would be some months before it was handling its capacity), which meant that derivatives of Bahian oil would soon reach national distributors. The CNP estimated potential production in 1950 at 20,000 b.p.d., and reserves at 44 million bbl.: Conselho Nacional do Petróleo, Relatório 1950 (Rio de Janeiro, 1951), pp. 9–11.

71 Wirth, Politics of Development, pp. 176–77.

72 The Petroleum Centre had gained public sympathy in part because of police persecution. On at least one occasion, the special police broke up, with considerable force, a public gathering sponsored by the Petroleum Centre. In 1948 the first "State Congress for the Defence of Petroleum and the National

Economy" was held in Rio de Janeiro, and its inaugural session took place at the end of the Avenida Rio Branco, Rio's main street. The crowd was dispersed and several participants were beaten: see Lôbo Carneiro's speech in the Chamber of Deputies, January 30, 1952, in Câmara dos Deputados, *Petróleo*, Vol. VI, pp. 70–71. Standard Oil of Brazil helped the nationalists' cause by sponsoring a campaign in the communications media in the hope of bringing down the Petroleum Statute and perhaps bringing about a more open oil law: "Parecer da Comissão de Indústria e Comércio da Câmara dos Deputados", *ibid.*, Vol. IV, p. 195. "Captain X" rose to the challenge and bitterly attacked "the trusts": Capitão X, "Comentários à Margem do Anteprojeto de Estatuto do Petróleo", *Revista do Clube Militar*, No. 103 (January 1950), pp. 4–12; *ibid.*, No. 107 (July, 1950), pp. 63–71; *ibid.*, No. 108 (August 1950), pp. 113–21. Other articles followed in the same vein.

73 Skidmore, *Politics in Brazil*, pp. 59, 65, 67, 73–79.

74 Vargas, *A Política Nacionalista do Petróleo no Brasil*, pp. 61–65.

Chapter 4

1 Thomas E. Skidmore, *Politics in Brazil, 1930–1964: An Experiment in Democracy* (New York: Oxford University Press, 1967), pp. 73–80; John D. Wirth, *The Politics of Brazilian Development, 1930–1954* (Stanford: Stanford University Press, 1970), pp. 184–88.

2 "Considerações sôbre a Guerra no Coreia", *Revista do Clube Militar* (July 1950), cited in Nelson Werneck Sodré, *História Militar do Brasil* (Rio de Janeiro: Editôra Civilização Brasileira, 1965), p. 312, and *Memorias de um Soldado* (Rio de Janeiro: Civilização Brasileira, 1967), pp. 305–6, 313.

3 Skidmore, *Politics in Brazil*, pp. 103–8.

4 J. Soares Pereira, "Depoimento", in Getúlio Vargas, *A Política Nacionalista do Petróleo no Brasil* (Rio de Janeiro: Livraria José Olympio Editôra, 1964), pp. 39–41; Alfredo Marques Vianna, "Apresentação", in *ibid.*, p. 19.

5 From Vargas's speech presenting his oil-company proposal to Congress: Vargas, *A Política Nacionalista do Petróleo no Brasil*, pp. 78–80.

6 *Os Fundamentos da "Petrobrás"* (Rio de Janeiro: Imprensa Nacional, 1952), pp. 13–15, 19, 30; Wirth, *Politics of Development*, p. 185; Skidmore, *Politics in Brazil*, pp. 82–100.

7 *New York Times*, May 7 and 30, 1951.

8 Lobo Carneiro, in Brasil, Congresso Federal, Câmara dos Deputados, Departamento dos Serviços de Taquigrafia, Directoria de Documentação e Publicidade, *Petróleo*, Vol. VI (Rio de Janeiro, 1956–59), pp. 70–71; *New York Times*, March 31, 1952.

9 Brasil, Ministério das Minas e Energia, Conselho National do Petróleo, *Relatório de 1951* (Rio de Janeiro, 1952), pp. 13–17.

10 Wirth, *Politics of Development*, p. 188; Mário Victor, *A Batalha do Petróleo Brasileiro* (Rio de Janeiro: Civilização Brasileira, 1970), pp. 292–93.

11 Vargas, *A Política Nacionalista do Petróleo no Brasil*, pp. 81–100; *New York Times*, December 7, 1951, and January 4, 1952.

12 Law 2004, in which form Petrobrás emerged from Congress late in 1953, may be found in Appendix III. Its only significant difference from the original bill is the exclusion of foreign investors (see Article 18).

13 Câmara dos Deputados, *Petróleo*, Vol. VI, p. 40. Rocha's bill was known as Project 1595/52.

14 *Ibid.*, pp. 34–37.

15 *Ibid.*, pp. 85, 121.

16 The PSP (Progressive Social Party) had recently emerged as the personal following of Adhemar de Barros, freewheeling and corrupt Governor of São Paulo, who had "delivered" his state to Vargas in the 1950 election.

17 Câmara dos Deputados, *Petróleo*, Vol. VI, pp. 52–53, 57.

18 *Ibid.*, pp. 157–58, 250.

19 Skidmore, *Politics in Brazil*, p.98; Odilon Braga, interview, in *O Jornal*, January 3, 1957; *Correio da Manhã*, October 16 and 20, 1954; personal interview, Glycon de Paiva, Rio de Janeiro, November 3, 1967.

20 Câmara dos Deputados, *Petróleo*, Vol. VII, p. 75.

21 *Ibid.*, pp. 195–96, 411, 449; J. Soares Pereira, *Última Hora*, January 6, 1951. Vargas had earlier consolidated deals with many PSD state organizations, which brought him a considerable share of his electoral support, despite the fact that the PSD had had its own candidate in the 1950 election: see Skidmore, *Politics*

in Brazil, p. 91, on the economic philosophies each major party tended to espouse.

22 Vargas, *A Política Nacionalista do Petróleo no Brasil*, p. 118.

23 Skidmore, *Politics in Brazil*, pp. 105–8; Sodré, *História Militar do Brasil*, p. 341; *O Cruzeiro* (June 7, 1952), p. 108.

24 Câmara dos Deputados, *Petróleo*, Vol. VI, pp. 70–71; *ibid.*, Vol. VII, pp. 266, 449, 451–52, 717–19.

25 *Ibid.*, Vol. VII, pp. 29–50.

26 João Café Filho, *Do Sindicato ao Catete: Memórias Políticas e Confissões Humanas* (Rio de Janeiro: Livraria José Olympio Editôra, 1966), Vol. II, p. 462.

27 Câmara dos Deputados, *Petróleo*, Vol. VII, pp. 57–58.

28 *Ibid.*, pp. 447–48, 490–92, 705–9.

29 Wirth, *Politics of Development*, pp. 199–200.

30 Câmara dos Deputados, *Petróleo*, Vol. VIII, pp. 5–9. There were five army generals present at the convention, among them Felicíssimo Cardoso, Artur Carnaúba, Antônio Henning, and Vicente de Paula Vasconcelos e Babaum; the navy was represented by Vice-Admiral Mandon.

31 *Ibid.*, Vol. VII, pp. 531, 602; *ibid.*, Vol. VIII, p. 125.

32 *Ibid.*, Vol. VIII, pp. 658, 679–81; *New York Times*, September 20, 1952.

33 Câmara dos Deputados, *Petróleo*, Vol. VIII, pp. 656–59.

34 *New York Times*, September 4, 1952.

35 Prof. Maurício Joppert, "O Petróleo no Brasil", *Revista do Clube de Engenharia* (July 1952), pp. 199–203. This is a reproduction of a speech Joppert gave in the Chamber of Deputies on June 18, 1952.

36 *Folha da Manhã* (Supplement), May 1953, p. 20. *O Estado de São Paulo* was of the same attitude on oil exploitation.

37 *New York Times*, August 11 and 18, 1952.

38 Skidmore, *Politics in Brazil*, pp. 115–16.

39 Conselho Nacional do Petróleo, *Relatório de 1952* (Rio de Janeiro, 1953), pp. 14–16; Brasil, Ministério das Minas e Energia, Conselho Nacional do Petróleo, *Resoluções do Plenário, 1945 a 1967*, Vol. I (mimeographed, Rio de Janeiro, n.d.), pp. 46–49.

40 Câmara dos Deputados, *Petróleo*, Vol. IX, pp. 95, 135, 146–48.

41 *Ibid.*, p. 153.

42 *Ibid.*, p. 100.
43 *Ibid.*, p. 351.
44 *Ibid.*, pp. 367–69.
45 Ibid., Vol. x, pp. 81, 91.
46 *Ibid.*, p. 5.
47 *Ibid.*, pp. 82–87, 91–95, 431–41.
48 The Commercial Association of Pôrto Alegre was the only such group to send a message to a deputy, and the association supported a company on the style of Volta Redonda: *ibid.*, Vol. VIII, p. 13.
49 *Ibid.*, Vol. x, pp. 177–79.
50 *Ibid.*, pp. 87–89, 99, 103.
51 *Ibid.*, Vol. XI, pp. 69, 72–74.
52 Brasil, Congresso Federal, Câmara dos Deputados, *Petróleo, Projetos 1516/51, 1517/51, 1595/52: Depoimentos, Pareceres e Votos* (Rio de Janeiro, 1952), pp. 75–83.
53 Câmara dos Deputados, *Petróleo*, Vol. XII, pp. 150, 175–83.
54 *Ibid.*, pp. 185, 187. For Daniel Faraco's views, see "Parecer da Comissão de Economia", Câmara dos Deputados, *Petróleo, Projetos 1516/51, 1517/51, 1595/52*, pp. 340–48.
55 "Parecer da Commisão Especial às Emendas do Senado", Câmara dos Deputados, *Petróleo*, Vol. XII, pp. 201–45.
56 *Ibid.*, pp. 290–95.
57 *Ibid.*, pp. 354–73.
58 *New York Times*, October 6, 1953.
59 Dahl Duff, "Brazil Turns Key in Closed Door", *Oil and Gas Journal* (November 9, 1953), p. 79; R. G. Walker, "Government Oil Monopoly Confirmed in Brazil", *World Petroleum*, Vol. XXIV, No. 12 (November 1953), pp. 118–22.
60 Dirceu Lino de Mattos, "O Petróleo no Brasil", in Etienne Dalemont, *O Petróleo* (São Paulo: Coleção Saber Atual, 1961), p. 160; J. Soares Pereira, "Depoimento", in Vargas, *A Política Nacionalista do Petróleo no Brasil*, p. 47; Café Filho, *Do Sindicato*, Vol. II, p. 463.
61 *New York Times*, January 1 and 25, 1954.
62 *Ibid.*, April 2 and July 9, 1954; *World Oil* (August 15, 1954), p. 140. The Cubatão refinery had been planned during the Dutra Administration.
63 *New York Times*, April 3, 1954; R. G. Walker, "Brazilian

Council Extends Exploration", *World Petroleum*, Vol. xxv, No. 7 (July 1954), p. 57; Conselho Nacional do Petróleo, *Relatório 1954*, (Rio de Janeiro, 1955), pp. 69–168.

64 Conselho Nacional do Petróleo, *Relatório 1953* (Rio de Janeiro, 1954), pp. 14–16, 231, 235, 237; *Relatório 1954*, p. 14; Sílvio Fróes Abreu, *Recursos Minerais do Brasil*, Vol. II, *Combustíveis Fósseis e Minérios Metálicos* (Rio de Janeiro: Ministério da Indústria e do Comércio, Instituto Nacional de Tecnologia, 1962), p. 181.

65 "Petrobrás como Banco da Subversão e Escola Prática de Corrupção", *Digesto Econômico*, Vol. xx, No. 176 (March–April 1964), p. 64. See also *New York Times*, June 19, 1954.

66 *New York Times*, May 11, 1954; *World Petroleum* (July 1954), p. 57.

67 Personal interview, Irnack Carvalho do Amaral, Rio de Janeiro, July 6, 1970.

68 Juarez Távora, "Situação Atual do Problema Petrolífera no Brasil: Elementos Extraídos da Conferência Realizada no E.S.G., em 4 de Junho de 1954", *Petróleo para o Brasil* (Rio de Janeiro: Livraria José Olympio Editôra, 1955), pp. 287, 299–301.

69 Skidmore, *Politics in Brazil*, pp. 122–42.

Chapter 5

1 *New York Times*, March 4, 1954.

2 *Ibid.*, March 12, 1954; Brasil, Ministério das Minas e Energia, Conselho Nacional do Petróleo, *Resoluções do Plenário, 1945 a 1967*, Vol. I (mimeographed, Rio de Janeiro, n.d.), pp. 71–72.

3 *A Noite*, January 3, 1955; personal interview, Irnack Carvalho do Amaral, Rio de Janeiro, June 29, 1970.

4 Glycon de Paiva, "Regime Legal e Produção de Petróleo", *Revista Brasileira de Economia*, Vol. VI, No. 2 (June 1952), pp. 58–60; Gral. Mário Poppe de Figueiredo, *Combustíveis Líquidos* (Rio de Janeiro: Imprensa Nacional, 1958), p. 22.

5 Glycon de Paiva, "Petrobrás como Banco da Subversão Nacional e Escola Prática de Corrupção", *Digesto Econômico* (March–April 1964), p. 64.

6 Thomas E. Skidmore, *Politics in Brazil, 1930–1964: An Experiment in Democracy* (New York: Oxford University Press, 1967), p. 143.

7 *New York Times*, September 21 and 28, 1954. The parent commercial federation and its various state subsidiaries remained at the centre of opposition to Petrobrás from its inception, as they had opposed its formation in messages to the Federal Senate. For example, in 1955, the new president of the Rio Commercial Association condemned "outmoded economic nationalism", which kept foreign capital from participation in development of Brazilian oil resources: *New York Times*, June 2, 1955.

8 *Correio da Manhã*, October 16–17 and 19, 1954.

9 See, for example, *O Jornal*, August 14–16, 1957.

10 *Diário de Notícias*, October 24–29, 1954. The story in the first issue carried the headline, "The Battle of Petrobrás".

11 Juarez Távora, "Quem Matou Vargas; Notas à Margem do Tema", *Manchete*, No. 822 (January 20, 1968), p. 128; *New York Times*, November 3, 1954; *Diário de Notícias*, October 29, 1954.

12 Eladio Marques, "A Verdade sôbre o Petróleo do Brasil", *O Observador Econômico e Financeiro*, No. 231 (May 1955), pp. 54–60; Skidmore, *Politics in Brazil*, pp. 145–46.

13 *New York Times*, November 3 and 6, 1954; João Café Filho, *Do Sindicato ao Catete: Memórias Políticas e Confissões Humanas* (Rio de Janeiro: Livraria José Olympio Editôra, 1966), Vol. II, pp. 466–68; Nelson Werneck Sodré, *História Militar do Brasil* (Rio de Janeiro, Editôra Civilização Brasileira, 1965), p. 359.

14 *New York Times*, March 12 and 15, 1955.

15 *Ibid.*, March 20, 1955; 4° Congresso Internacional de Petróleo, *Brasil: O Petróleo da Amazônia* (Rome, 1955), no page numbers.

16 Glycon de Paiva, "O Petróleo de Nova Olinda", *Carta Mensal*, Vol. I, No. 3 (June 1955), pp. 42–44.

17 Robert J. Alexander, *The Bolivian National Revolution* (New Brunswick, N.J.: Rutgers University Press, 1958), pp. 160–62; Cornelius H. Zondag, *The Bolivian Economy, 1952–1965: The Revolution and Its Aftermath* (New York: Frederick A. Praeger, 1966), pp. 112–13; Mário Leão Ludolf, *Da Conveniência da Exploração do Petróleo Boliviano* (Rio de Janeiro: Confederação Nacional da Indústria, Conselho Econômico, 1959), p. 11; [Refinaria e Exploração do Petróleo União], *Petróleo Boliviano e Mercado*

Brasileiro (São Paulo, 1956), pp. 34–35; Café Filho, *Do Sindicato*, Vol. II, pp. 445, 447–48.

18 *New York Times*, April 1, 1955.

19 Jarbas Maranhão, "Petróleo no Senado", *O Observador Econômico e Financeiro*, No. 246 (August 1956), pp. 56–60; Senador Jarbas Maranhão, "Vitória da Petrobrás no Senado", *Revista do Clube Militar*, No. 143 (July–September 1956), pp. 75–80.

20 *World Petroleum*, Vol. XXVII, No. 9 (August 1956), p. 88.

21 *New York Times*, August 23, 1956; personal interview, Irnack Carvalho do Amaral, Rio de Janeiro, October 31, 1967. Nunes's appointment was probably the most political of the three to that time; a member of the PSD, he had been treasurer of Kubitschek's presidential campaign.

22 *O Semanário*, July 12–19, 1956.

23 *Ibid.*, June 6–13, 1956.

24 Zondag, *Bolivian Economy*, pp. 113–14; Alexander, *Bolivian National Revolution*, pp. 164–66; Café Filho, *Do Sindicato*, Vol. II, pp. 450–52; *Petróleo Boliviano e Mercado Brasileiro*, pp. 34–35.

25 Skidmore, *Politics in Brazil*, pp. 166–74.

26 *O Semanário*, May 24–30, 1957.

27 *Última Hora*, August 6 and 9, 1957.

28 *Ibid.*, August 30, 1958; Skidmore, *Politics in Brazil*, pp. 143–58.

29 *Última Hora*, June 24, 1957; *New York Times*, June 30, 1957.

30 *New York Times*, April 20, 1957.

31 *Última Hora*, May 2, 1957.

32 *O Semanário*, May 9–16, 1957.

33 *Última Hora*, August 24, 1958.

34 *Ibid.*, August 19, 1957.

35 *Última Hora*, December 31, 1957; *World Oil* (August 15, 1958), p. 156; *New York Times*, July 18, 1957. Average consumption for 1956 was 200,000 b.p.d.; for 1957, 210,000 b.p.d. Average production for 1956, 11,000 b.p.d.; for 1957, 27,000 b.p.d. Total imports in 1956: crude oil from the Middle East and Venezuela, $105,900,000; refined products, $172,600,800.

36 *New York Times*, July 2, 1954, January 5, 1955, and January 6, 1957; Conselho Nacional do Petróleo, *Resoluções do Plenário*, Vol.

I, pp. 119–20. The Cubatão refinery had been part of the SALTE Plan during the Dutra Administration.

37 *New York Times,* February 9 and November 19, 1957, and May 3 and August 24, 1958; "Brazil", *World Petroleum Report,* Vol. III (1957), p. 52.

38 "Não Manda mas Pega", *O Jornal,* October 27, 1957.

39 "Petrobrás em Perigo", editorial, *Diário de Notícias,* April 8, 1958. The newspaper normally supported Petrobrás.

40 *Ibid.,* April 8, 1958. The letter was dated November 11, 1957. In a personal interview with the author October 31, 1967, Dr. Irnack Carvalho do Amaral confirmed the charges made by Figueiredo in his letter, and said he had resigned with his fellow director because he felt that, since Nunes continually bypassed the directorate, he could not be of use to the company.

41 *Diário de Notícias,* April 11 and 15, 1958. Nunes later published a defence of his record: Janari Gentil Nunes, *Defesa dos Programas da Petrobrás* (Rio de Janeiro, 1959).

42 *Diário de Notícias,* April 20 and May 23, 1958. The Mataripe refinery in the Recôncavo at that time processed 5,000 b.p.d. of Bahia production, but subsequently was expanded to 75,000 b.p.d. capacity. Petrobrás later converted another refinery to process Brazilian crude.

43 *Ibid.,* April 24, 1958.

44 Nunes had his testimony before the Parliamentary Inquiry Committee broadcast to the nation over the radio; in one of the last shots in its campaign, *Diário de Notícias* asked whether Petrobrás was paying for this publicity (May 31, 1958).

45 *Diário de Notícias,* May 7, 1958; *New York Times,* August 21, 1958; W. J. Levy, Inc., *The Search for Oil in Developing Countries: A Problem of Scarce Resources and Its Implications for State and Private Enterprise,* prepared at the request of the International Bank for Reconstruction and Development (New York, 1960), p. 80. Bittencourt also criticized Petrobrás for improper conservation practices.

46 *Última Hora,* November 4 and 10, 1958.

47 *Ibid.,* December 9, 1958; *New York Times,* December 10, 1958; Glycon de Paiva, *Digesto Econômico* (March–April 1964), p. 65.

48 Hélio Beltrão, "Seis Equívocos Fundamentais sôbre a Petrobrás", *Última Hora,* August 22, 1958. The newspaper finished

the essay on August 23; it was subsequently published in pamphlet form.

49 Hélio Jaguaribe, *O Nacionalismo na Atualidade Brasileira* (Rio de Janeiro: Instituto Superior de Estudos Brasileiros, 1958), pp. 156–57. He wrote his study at the time when Janari Nunes was president of Petrobrás and he remarked on the unfortunate trend being followed by the company at that time.

50 *Última Hora,* July 16–22, September 4–6, October 1, and November 5, 1957; *O Estado de São Paulo,* October 4, 1957; *New York Times,* October 6 and November 2, 1957; *O Semanário,* February 21–28, 1958.

51 Lucio Glauco Torres, "Petróleo da Bolívia", *O Observador Econômico e Financeiro,* No. 279 (May 1959), p. 52

52 "Bolívia", *World Oil,* Vol. 149, No. 3 (1959), pp. 131–32; Café Filho, *Do Sindicato,* Vol. II, pp. 452–53; *Petróleo Boliviano e Mercado Brasileiro,* pp. 13–14.

53 *Última Hora,* January 30–31, February 7 and April 2, 1958; Torres, "Petróleo da Bolívia", p. 52.

54 *Última Hora,* September 13, 1958; *New York Times,* September 13, 1958.

55 *Última Hora,* September 23, 1958.

56 *Ibid.,* October 22 and November 6, 1958.

57 Skidmore, *Politics in Brazil,* pp. 174–80.

58 *Última Hora,* January 12–14, 1959; Torres, "Petróleo da Bolívia", p. 52.

59 Apart from the newspapers cited above, particularly for late January 1959, see Lourival Fontes, *Política, Petróleo e População* (Rio de Janeiro: Livraria José Olympio Editôra, 1958), pp. 55–56; Joel Silveira and Lourival Coutinho, *História de uma Conspiração (Bolívia, Brasil e Petróleo)* (Rio de Janeiro: Editôra Coelho Branco, 1959); Anderson O. Mascarenhas, *Roboré, um Torpedo contra a Petrobrás* (São Paulo: Editôra Fulgor, 1959); Gabriel de Rezende Passos, *Estudo sôbre o Acordo de Roboré* (São Paulo: Editôra Fulgor, 1960). A good reply to these cases against the agreement is Olympio Guilherme, *A Verdade sôbre Roboré* (Rio de Janeiro: Livraria Freitas Bastos, 1960).

60 Café Filho, *Do Sindicato,* Vol. II, pp. 452–53; personal interview, Lauriston Pessôa Monteiro, Rio de Janeiro, November 6, 1970. Dr. Monteiro was then a superintendent of União

Brasil–Bolívia de Petróleo S.A. Petrobrás took over the company in 1973.

61 *Diário de Notícias,* April 2, 1958. Rollie E. Poppino, *International Communism in Latin America: A History of the Movement 1917–1963* (New York: Free Press, 1964), p. 169, notes that in 1957 delegates to Moscow from Latin America were instructed "to emphasize nationalism, [and] to identify themselves with the legitimate aspirations of the Latin American people."

62 *Última Hora,* January 17, 1957, and June 27, 1958.

63 *Ibid.,* August 4, 1958.

64 *Ibid.,* July 17, 22, and 24, 1958.

65 *New York Times,* August 10, 1958.

66 *Última Hora,* August 1 and 6–8, 1958.

67 *Ibid.,* August 6, 1958; *ibid.,* August–December 1958, *passim.* On August 7, the paper quoted General Lott to the effect that he saw "no reason to change the orientation we have followed for the exploitation of our petroleum."

68 *O Jornal,* July 26–29, 1958. One of the newspaper's columnists, Abelardo da Cunha, chastised Brazil for its "stupidity" in continuing to espouse a policy of "O petróleo é nosso" in the light of events in Argentina. He pointed out that Brazil was at the mercy of foreign events such as crises in the Middle East because of its blindness. He seemed to imply that Brazil had oil "under its feet", yet it continued to import vast amounts. He said this was the case with Argentina, but it had wakened to reality (*O Jornal,* July 27). This is true, but Argentina knew it had vast reserves while Brazil did not; here was another instance of the "childlike faith" Brazilians held in the oil potential of their country that Odilon Braga had talked about over twenty years before.

69 *New York Times,* March 10, 1959; Levy, *The Search for Oil,* p. 74.

70 *Jornal do Brasil,* November 21, 1959.

71 Roberto de Oliveira Campos, "As Três Falácias do Momento Brasileiro", *Digesto Econômico,* Vol. XIII, No. 133 (January–February 1957), pp. 49–66.

72 See n. 48.

73 *Folha da Manhã,* July 3 and 24, 1958.

74 *Última Hora,* August 15, 1958.

75 Skidmore, *Politics in Brazil*, p. 180.
76 *Última Hora*, April 23–28, 1959.
77 Walter K. Link, "What are the Chances for Oil in the Sedimentary Basins of Brazil?", *Oil and Gas Journal*, Vol. LVII, No. 47 (November 16, 1959), pp. 289–94. There was no comment in the Brazilian press on this article. One Brazilian, Glycon de Paiva, did see Link's paper and used his findings as the basis of an article on Brazil's oil potentiality: Glycon de Paiva, "Reavaliação das Possibilidades Petrolíferas do Brasil (1960)", *Carta Mensal*, Vol. VI, No. 63 (June 1960), pp. 9–24.
78 "Link Reports on Brazil", *World Petroleum*, Vol. XXXII, No. 3 (March 1961), pp. 36–38; "Brazilian Oil Picture: Hopeless, says Topflight U.S. Geologist", *Oil and Gas Journal* (January 9, 1961), pp. 58–59; personal letter, Walter K. Link to the author, November 15, 1965, in the possession of the author; personal interview, Walter K. Link, LaPorte, Indiana, September 16, 1966. Link signed a contract for five years in 1954, and agreed to stay on one more year after his contract expired.
79 *Oil and Gas Journal* (January 9, 1961), p. 58; *Última Hora*, November 24, 1960; Gabriel Passos, "Nacionalismo e Riquezas Minerais", *Temas Nacionalistas* (São Paulo, 1961), pp. 90, 94. Link made his first public statement in Brazil in answer to Passos's attack. He declared he had been exclusively a technician while with Standard Oil and had never taken part in the direction or policy of the company; furthermore, he had had no further contact with the personnel of Standard Oil since resigning to take the job at Petrobrás. He considered Passos's charge that he had been confused with his brother "an insult to Petrobrás". The system of selecting a drilling site ruled out any possiblity of sabotage; too many people had to approve the site: *O Globo*, November 28, 1960.
80 *Última Hora*, November 26, 1960. Brigadier Henrique Fleiuss, president of the CNP, echoed Sardenberg's statement: *ibid.*, November 29, 1960.
81 Personal interview, Irnack Carvalho do Amaral, Rio de Janeiro, July 3, 1970.
82 *Última Hora*, December 3, 1960.
83 *Ibid.*, November 24 and 29, 1960.
84 Gen. Tácito Freitas, *Petróleo apesar de Mr. Link* (Rio de

Janeiro: Edições Gernasa, 1964), pp. 12–15; *Última Hora*, December 23, 1960. The Cr$140 million probably referred to Link's salary, reportedly $100,000 per year.
85 Levy, *The Search for Oil*, pp. 57–82.
86 *Ibid.*, pp. 60–71 (Table 1 is taken from p. 61).
87 Figueiredo, *Combustíveis Líquidos*, p. 27; "Brazil", *World Petroleum Report*, Vol. IV (January 15, 1958), p. 100.
88 Levy, *The Search for Oil*, p. 64.
89 *Última Hora*, July 1, 1960. Assuredly, Brazil would be able to get a much better price for all concerned if one entity had all imports as a bargaining-point; however, the figure Lima Sobrinho cited cannot be evaluated due to lack of data about import costs.
90 *Última Hora*, December 8, 1959. No further details of the contract were made available.
91 Levy, *The Search for Oil*, pp. 72–74; *Oil and Gas Journal* (January 9, 1961), pp. 58–59.

Chapter 6

1 Lott had been prominent as a firm endorser of constitutional legality and a nationalist of a vaguely radical sort since 1955. Unfortunately for the PSD-PTB leaders, he was politically inexperienced and lacking in personal appeal, a weak candidate: Thomas E. Skidmore, *Politics in Brazil, 1930–1964: An Experiment in Democracy* (New York: Oxford University Press, 1967), pp. 189–90.

2 *Jornal do Brasil*, December 4, 1960. There was a further controversy over how to supply the refinery, particularly over a proposal to by-pass existing rail transportation and build a pipeline: *O Estado de São Paulo*, January 12–13, 1961; J. Soares Pereira, "Do Ponto de Vista Nacional", *Última Hora*, March 17, 1961; "Oleoduto Rio–Belo Horizonte: Será Favorável à Economia Nacional?", *Brasil de Hoje* (São Paulo), No. 78 (September 1961), p. 23.

3 *O Estado de São Paulo*, December 10, 1960; Alfredo Marques Vianna, "Apresentação", in Getúlio Vargas, *A Política Nacionalista do Petróleo no Brasil* (Rio de Janeiro: Livraria José Olympio Editôra, 1964), pp. 23–24; Eugênio Gudin, *Análise de Problemas Brasileiros, 1958–1964* (Rio de Janeiro: Agir, 1965), p. 380;

Eugênio Gudin, "Petrobrás: Desperdício e Incapacidade", *Digesto Econômico*, Vol. XVII, No. 158 (March–April 1961), p. 99.

4 Gabriel Passos, in *Temas Nacionalistas* (São Paulo, 1961), pp. 10–11; "Nacionalismo e Refinarias de Petróleo", *O Estado de São Paulo*, December 3, 1960.

5 *New York Times*, February 23, 1961.

6 Glycon de Paiva, *Digesto Econômico* (March–April 1964), p. 66; *Jornal do Brasil*, May 27, 1962.

7 *Última Hora*, March 4, 1961.

8 *Ibid.*, April 14, 19, and 27, 1961; *New York Times*, April 16, 1961.

9 See Chapter 5, n.91, for Link's estimate. *World Petroleum Report*, Vol. VII (1961), p. 242.

10 *World Petroleum Report*, Vol. VII (1961), p. 243. Data on such exchanges were not available to the author; it is thus impossible to draw conclusions as to whether they were advantageous to Brazil or not. Moreover, one cannot generalize that heavy crude —which, as has been mentioned, yields less gasoline per barrel than lighter varieties—is consistently or even usually worth less than light crude. Current market conditions, influenced by a multitude of factors, determine the value of different grades of crude oil at different times. It is therefore quite possible that Brazil was able to trade its crude, barrel for barrel, for the type it wanted, or even at a profit; it is, of course, also possible that it had to sell at a loss.

About all one can conclude is that Brazil's refinery complex was not as large or as flexible as that of "the trusts" with whom she was exchanging crudes. The Brazilian private refineries had been built to process foreign crude; Petrobrás's refinery at Cubatão, São Paulo, had been planned in the previous decade under the abortive SALTE Plan, to process foreign crude (in order to generate revenue for exploration). The refining sector had not yet come adequately to grips with national crude-oil production.

11 *Última Hora*, June 28, 1961. General Felicíssimo Cardoso was still president of the Petroleum Centre.

12 *Ibid.*, June 1961.

13 Personal interview, Irnack Carvalho do Amaral, Rio de Janeiro, July 6, 1970.

14 *Última Hora*, July 26, August 3, and September 19, 1961.

15 Interested readers should start with Skidmore, *Politics in Brazil*, pp. 200–11; the relevant footnotes give a bibliography (to 1966) of the crisis.
16 *Última Hora*, December 5, 1961.
17 *O Semanário*, December 10–17, 1961.
18 *Última Hora*, December 28, 1961.
19 *Ibid.*, January 2, 1962.
20 *Ibid.*, January 6, 1962. Deputy Clemens Sampaio later said Passos was inconsistent, since shortly after calling Amaral an *entreguista* he appointed him to the top position of Petrobrás: *O Globo*, December 9, 1962. Amaral had been a member of the executive directorate of Petrobrás when appointed to fill the vacancy left by Barroso.
21 *O Globo*, January 8, 1962.
22 *Última Hora*, January 9, 1962.
23 João Pinheiro Neto, "Com Quem a Verdade?", *Última Hora*, January 12, 1962.
24 *Última Hora*, January 10–12, 1962.
25 *O Globo*, January 15, 1962; *Última Hora*, January 15, 1962. Passos apparently preferred Eduardo Sobral for president of Petrobrás, but labour pressure had made him change his mind.
26 *Última Hora*, January 15, 1962.
27 Glycon de Paiva, *Digesto Econômico* (March–April 1964), p. 66; *O Globo*, January 16, 1962; personal interview, Irnack Carvalho do Amaral, Rio de Janeiro, October 31, 1967.
28 *O Globo*, January 16, 1962.
29 *Ibid.*, January 16, 1962.
30 *Última Hora*, January 18, 1962.
31 *O Globo*, January 19, 1962.
32 Gudin, *Análise de Problemas Brasileiros*, p. 381 (an article written February 26, 1962).
33 *O Globo*, December 28, 1961.
34 Alfred Stepan, *The Military in Politics: Changing Patterns in Brazil* (Princeton: Princeton University Press, 1970), pp. 85–121.
35 *Diário de Notícias*, September 23, 1961; *Jornal do Comércio*, August 27, 1961; *Diário Popular* (São Paulo), May 9, 1963, said 68 per cent of Duque de Caxias refinery was Brazilian-made.
36 *Última Hora*, December 29, 1961, September 19, 1962, and June 26 and September 4, 1964; *World Petroleum* (February 1965),

p. 29; Francisco Mangabeira, "A Petrobrás e a Expansão Industrial Brasileira", *Desenvolvimento e Conjuntura*, Vol. VI, No. 11 (November 1962), pp. 105–7; "Petrobrás Não Perdeu o Ritmo", *Visão*, November 20, 1964.

37 *Petrobrás*, Vol. VII, No. 179 (January 1961), pp. 4–5.

38 José Jofilly (PSD, later PSB, leader of the FPN), in Brasil, Congresso Federal, *Anais da Câmara dos Deputados*, 3rd Session (February 6, 1962), pp. 198–205.

39 *Última Hora*, February 7, 1962.

40 *Ibid.*, March 2, 1962. It should be emphasized that this well was a wildcat, just as that of Nova Olinda had been. Petrobrás kept Tucano in reserve until recently and did little exploration there until late in the decade, preferring to continue to develop the Recôncavo. Tucano is inland, and development of its reserves is more costly than in the Recôncavo because of distance.

41 *Ibid.*, March 15, 1962; *New York Times*, March 16, 1962.

42 *Última Hora*, February 12, 1962; *New York Times*, January 10, 1962.

43 *Última Hora*, February 21, 1962.

44 *Ibid.*, March 8, 1962.

45 *Ibid.*, March 12, 1962; *New York Times*, March 18, 1962. No further mention of such loans was made, and it appears from subsequent annual reports of Petrobrás that it went ahead on its own. The operation is still in the pilot stage, and no concrete results appear to have been made public.

46 *Última Hora*, December 2, 1961, and May 3, 1962; *O Jornal*, May 1, 1962; *Jornal do Brasil*, May 27, 1962.

47 *Anais da Câmara dos Deputados*, 42nd Session (May 7, 1962), pp. 458–60; *Última Hora*, May 9, 1962; *Jornal do Brasil*, May 27, 1962.

48 *Última Hora*, May 15 and 19, 1962; *Jornal do Brasil*, January 22, 1964.

49 Pedro Muller, "Petrobrás em Crise", *Jornal do Brasil*, May 24 and 27, 1962; *New York Times*, May 27, 1962.

50 "Dos Técnicos da Petrobrás à Nação", *Jornal do Brasil*, May 26, 1962.

51 *Ibid.*, May 27 and 29 and June 1–8, 13, 20–21, 24, and 27, 1962.

52 *Diário de Notícias*, August 4, 1962; *O Jornal*, August 7, 1962.

53 Petróleo Brasileiro S.A., *Relatório de Atividades em 1962* (Rio de Janeiro, 1963), cited in *Jornal do Comércio*, March 9, 1963.

54 *Jornal do Brasil*, September 23, 1962.

55 *Última Hora*, September 26, 1962.

56 *O Jornal*, December 7, 1962; *New York Times*, January 31, 1963; *Última Hora*, January 31, 1963; R. G. Walker, "Brazil Pushes to Expand Production", *World Petroleum*, Vol. xxxiv, No. 5 (May 1963), p. 34.

57 Skidmore, *Politics in Brazil*, p. 230; *Última Hora*, January 10 and 14–February 1, 1963.

58 *Última Hora*, January 7, 1963; Brasil, Ministério das Minas e Energia, Conselho Nacional do Petróleo, *Resoluções do Plenário, 1945 a 1967*, Vol. ii (mimeographed, Rio de Janeiro, n.d.), pp. 71–73.

59 *Última Hora*, August 9, 1961.

60 *New York Times*, January 16, 1963; Skidmore, *Politics in Brazil*, pp. 236–43.

61 *New York Times*, June 8, 1963; *Última Hora*, June 10, 1963; *Jornal do Brasil*, June 7, 1963.

62 *Jornal do Brasil*, June 7, 9, 11, and 13, 1963.

63 Skidmore, *Politics in Brazil*, pp. 253–84; Stepan, *The Military in Politics*, pp. 123–212; Ronald M. Schneider, *The Political System of Brazil: Emergence of a "Modernizing" Authoritarian Regime, 1964–1970* (New York: Columbia University Press, 1971), pp. 73–107.

64 *Última Hora*, June 11 and 12, 1963; *Jornal do Brasil*, June 12 and 13, 1963.

65 *Última Hora*, July 30, 1963.

66 *Anais da Câmara dos Deputados*, 118th Session (July 30, 1963), pp. 353–54.

67 *Ibid.*, 119th Session (July 31, 1963), pp. 510–11.

68 *Jornal do Brasil*, June 6 and August 20, 1963; *New York Times*, December 22, 1963; Gen. Tácito Freitas, *Petróleo apesar de Mr. Link* (Rio de Janeiro: Edições Gernasa, 1964), p. 199.

69 *Jornal do Brasil*, August 22, 1963.

70 *Ibid.*, August 17 and 22, 1963; Skidmore, *Politics in Brazil*, pp. 244–45, 247.

71 *Jornal do Brasil*, August 24, 1963.

72 *Última Hora*, August 19, 1963. The newspaper was then

tending toward the radical camp led by Brizola and away from the more moderate left.

73 *Ibid.*, September 9, 1963.

74 *Última Hora*, November 12, 1963; *New York Times*, November 7, 1963.

75 *New York Times*, November 14, 1963; *Última Hora*, November 13, 1963.

76 *Última Hora*, November 18, 1963.

77 *New York Times*, December 25, 1963; *Jornal do Brasil*, December 24 and 27, 1963; *Petroleum Intelligence Weekly*, quoted in *Diário Popular* (São Paulo), May 31, 1965.

78 *Jornal do Brasil*, August 25 and 27, 1963.

79 *New York Times*, October 14, 1963; *Última Hora*, October 15, 1963.

80 Tácito Freitas, writing in *O Semanário*, November–December 1963, reproduced in *Petróleo apesar de Mr. Link*, pp. 80, 83, 119–134.

81 Personal interviews, Irnack Carvalho do Amaral, Rio de Janeiro, October 31, 1967, and July 9, 1970.

82 *Jornal do Brasil*, October 10, 1963; *New York Times*, October 14, 1963. Tagiev and Bakirov were interviewed by a Soviet journal for Latin-American consumption upon their return to Russia. Asked to comment upon Link's letter, the technicians said, "We desire that time decides who is right," *Tiempos Nuevos*, November 19, 1963, reproduced in Freitas, *Petróleo apesar de Mr. Link*, pp. 143–44.

83 *Última Hora*, January 25, 1964.

84 *Jornal do Brasil*, January 26, 1964.

85 *Última Hora*, January 27, 1964; *New York Times*, January 28, 1964.

86 *Última Hora*, January 28, 1964; *New York Times*, January 29, 1964; *Jornal do Brasil*, January 28–31, 1964.

87 *Última Hora*, January 28, 1964; *Jornal do Brasil*, January 28, 1964.

88 *Jornal do Brasil*, January 30, 1964; *Última Hora*, January 31, 1964.

89 *Jornal do Brasil*, January 30 and February 4–5, 1964; *Última Hora*, February 1 and 4, 1964. The army *coup* in April stopped the work of both committees; neither issued a report.

90 Glycon de Paiva, *Digesto Econômico* (March–April 1964), pp. 67–68. The *New York Times* had said, in its issue of December 25, 1963, that Petrobrás owed "more than 100 million dollars" to suppliers of crude oil.

91 *Última Hora*, February 21 and March 7, 1964.

92 Skidmore, *Politics in Brazil*, pp. 253–302, particularly pp. 286–90; *New York Times*, March 14, 1964; *Última Hora*, March 14, 1964. The latter newspaper, which had come out so dramatically in opposition to the move a few months earlier, made absolutely no comment, either way, on the expropriation in the days that followed. See also above, pp. 52–5 and n.35, for a brief breakdown of political camps.

93 *Última Hora*, March 21, 1964.

94 Skidmore, *Politics in Brazil*, pp. 301–2; *New York Times*, April 2, 1964. Goulart echoed Vargas, charging that the wave of propaganda against him had been financed partly by the petroleum "trusts".

95 See Stepan, *The Military in Politics*, Part III, particularly pp. 188–212, for an excellent analysis of the *coup*.

Chapter 7

1 At the end of 1969 the federal government held 78.3 per cent of Petrobrás's common stock, subscribing 75.6 per cent of the company's share capital. Private individuals held but 7.7 per cent of common stock, with the remainder in the hands of states, municipalities, and "public agencies": Petróleo Brasileiro S.A., *Relatório de Atividades 1969* (Rio de Janeiro, 1970), p. 53. See Table 2 for the sources of Petrobrás's funds in 1969.

2 Charles R. Dechert, *Ente Nazionale Idrocarburi: Profile of a State Corporation* (Leiden, 1963), pp. 61–73, 100–8.

3 No fewer than eight men were president of Petrobrás between May 1954 and April 1964, and holders of the three directorate positions changed fifteen times in the same period: Ilmar Penna Marinho, Jr., *Petróleo: Soberania e Desenvolvimento* (Rio de Janeiro: Edições Bloch, 1970), pp. 414–17.

4 John D. Wirth, *The Politics of Brazilian Development, 1930–1954* (Stanford: Stanford University Press, 1970), pp. 126–29.

5 *Ibid.*, p. 170; Judith Tendler, *Electric Power in Brazil: Entrepre-*

neurship in the Public Sector (Cambridge: Harvard University Press, 1968), pp. 2, 63.

6 Tendler, *Electric Power in Brazil,* pp. 43–55, 79, 213.

7 *New York Times,* April 3, 1964.

8 *Última Hora,* April 4, 1964.

9 *Ibid.,* April 8, 1964; *Jornal do Brasil,* April 7, 1964.

10 *Jornal do Brasil,* April 4, 7–8, 10, 14, and 16–17, 1964.

11 *Ibid.,* April 23, 25, and 28–29 and May 5, 1964; *New York Times,* July 3, 1965.

12 *Jornal do Brasil,* May 5, 1964; *Última Hora,* May 5, 11, and 12, 1964; *New York Times,* May 10 and 13, 1964.

13 *Jornal do Brasil,* May 16 and 23, 1964.

14 *Última Hora,* June 12, 1964.

15 *Jornal do Brasil,* June 16, 1964. The colonel charged that the magazine had cost Cr$36 million ($36,000) per month to publish. He produced documents to support his charges.

16 *Ibid.,* July 14 and 22, 1964; personal interview, Irnack Carvalho do Amaral, Rio de Janeiro, October 31, 1967.

17 *Jornal do Brasil,* June 21, 1964.

18 *Ibid.,* August 2, 1964. The President of Brazil was turning the sod for the Rio–Belo Horizonte pipeline.

19 *New York Times,* July 3, September 12 and 28, and October 13, 1965.

20 "Brazil: the Petrochemicals Industry", *B.O.L.S.A. Review,* Vol. IV (June 1970), pp. 312–17.

21 *New York Times,* December 9 and 27, 1964; "Ademar de Queirós a O Globo: Carmópolis é o Maior Campo de Petróleo já Encontrado no País", *O Globo,* January 6, 1965; J. E. Rassmuss, "New Miranga Field to Help Meet Brazil's Need for Oil", *World Oil* (December 1966), p. 97; Brasil, Ministério das Minas e Energia, *Panorama do Setor Mineral Brasileiro* (October 1966), p. 30.

22 *Petroleum Press Service* (1968), p. 136; *ibid.* (1969), pp. 179–81; *International Petroleum Encyclopedia, 1970* (Tulsa: Petroleum Publishing Company, 1969), p. 182; Alvaro Franco, "Brazil Plans Wildcat Assault Offshore", *Oil and Gas Journal,* Vol. 65 (March 13, 1967), p. 86.

23 Rassmuss, "New Miranga Field", p. 97.

24 *Jornal do Brasil,* December 30, 1966; "Barreirinhas e 150 Mil Barris São os Maiores Tentos de 66", *Petrobrás* (January–Feb-

ruary 1967), pp. 9–10; personal interviews, Irnack Carvalho do Amaral, Rio de Janeiro, October 31, 1967, and June 29, 1970; Ministério das Minas e Energia, *Panorama do Setor Mineral Brasileiro*, p. 29; *World Oil* (January 1967), pp. 104–6.

25 *Jornal do Brasil*, December 30, 1966; *Petrobrás* (January–February 1967), pp. 7–8.

26 "Petrobrás at the Crossroads", *Petroleum Press Service* (April 1971), p. 139. The yield from the Recôncavo in 1970 was 135,000 b.p.d.

27 *Ibid.* (March 1971), p. 102. The shelf's production has increased; in the first quarter of 1974 it contributed 10.5 per cent of the national total of 180,000 b.p.d.: *Jornal do Brasil*, June 19, 1974.

28 Ministério das Minas e Energia, *Panorama do Setor Mineral Brasileiro*, p. 26.

29 "Petrobrás at the Crossroads", p. 139.

30 *Petrobrás*, No. 259 (January–February 1973), p. 42.

31 *Ibid.*, No. 248 (March–April 1971), pp. 34–38; *ibid.*, No. 255 (May–June 1972), pp. 12–15; "Petrobrás at the Crossroads", pp. 139–41.

32 *Jornal do Brasil*, April 16, 1967.

33 Petróleo Brasileiro S.A., *Exercício de 1966* (Rio de Janeiro, February 2, 1967); *Jornal do Brasil*, July 9, 1967; *Petrobrás* (March–April, 1971), p. 38.

34 "Petrobrás at the Crossroads", p. 141.

35 "200 Largest Industrials outside U.S. in terms of Sales", *Fortune*, Vol. LXXIV, No. 3 (August 1966), p. 149; *ibid.*, Vol. LXXXIV, No. 2 (August 1971), p. 152.

36 *Jornal do Brasil*, April 6, 1967.

37 Petróleo Brasileiro S.A., *Relatório de Atividades 1969*, pp. 53–70.

38 *Petrobrás*, No. 247 (January–February 1971), p. 5, published figures showing that Brazilians paid less for their gasoline (*c.* 40¢ per U.S. gallon) than motorists in thirteen other countries of Europe and South America.

39 Petróleo Brasileiro, S.A., *Relatório de Atividades 1969*, p. 55.

40 *O Globo*, August 13, 1965.

41 *Jornal do Brasil*, October 8 and 10, 1967. Consumption was, in fact, *c.* 800,000 b.p.d. in 1974.

42 *Petrobrás*, No. 248 (November–December 1970), p. 40.

43 See Introduction, pp. 2-3, for a brief discussion of this point.

44 I am indebted for this insight to Alfred Stepan's book, *The Military in Politics: Changing Patterns in Brazil* (Princeton: Princeton University Press, 1970). Stepan has clearly disproven the thesis that the Brazilian military is or was "outside" politics; my research merely confirms his thesis, with respect to the oil question, particularly after the Second World War.

BIBLIOGRAPHY

Official Publications

BELTRÃO, HÉLIO M. P. *Plano de Organização dos Serviços Básicos da Petrobrás.* Rio de Janeiro: Imprensa Nacional, n.d.

Boletim Técnico da Petrobrás. Rio de Janeiro, 1957—.

[Braga, Odilon]. *Ante-projeto do Estatuto do Petróleo com o Parecer do Relator.* Rio de Janeiro: Imprensa Nacional, 1948.

_____. *Bases para a Inquérito sôbre o Petróleo.* Rio de Janeiro: Imprensa Nacional, 1936.

Brasil. Congresso Federal. *Anais da Câmara dos Deputados.*

_____. Câmara dos Deputados. *Petróleo, Projetos 1516/51, 1517/51, 1595/52: Depoimentos, Pareceres e Votos.* Rio de Janeiro, 1952.

_____. Câmara dos Deputados. Comissão de Constituição e Justiça. *Estatuto do Petróleo. Parecer do Relator, Deputado Benedicto Costa Netto.* Rio de Janeiro, 1948.

_____. Câmara dos Deputados. Departamento dos Serviços de Taquigrafia. Diretoria de Documentação e Publicidade. *Petróleo.* 12 Vols. Rio de Janeiro. 1956–59.

Brasil. Ministério da Agricultura. Departamento Nacional da Produção Mineral. Divisão de Fomento da Produção Mineral. *Relatório da Diretoria, 1939,* por Octávio Barbosa. Rio de Janeiro, 1940.

_____. Juarez Távora. *O Ministro da Agricultura perante a Assembleia Nacional Constituinte.* Rio de Janeiro, 1934.

Brasil, Ministerio da Agricultura, Industria e Commercio. Serviço Geologico e Mineralogico do Brasil. Euzebio Paulo de Oliveira, Director. *Relatório Annual do Director, Anno 1928.* Rio de Janeiro, 1929.

Brasil, Ministério das Minas e Energia. *Panorama do Setor Mineral Brasileiro.* October 1966.

_____. Conselho Nacional do Petróleo. *Legislação do Petróleo*. Rio de Janeiro: Britânia Editôra, 1964.

_____. Conselho Nacional do Petróleo. *Relatório*. Rio de Janeiro, 1945–.

_____. Conselho Nacional do Petróleo. *Relatório, 1° triênio, 1938–41*, and *Anexos*, Vols. 1 and 2. Typewritten. Rio de Janeiro, 1941.

_____. Conselho Nacional do Petróleo. *Resoluções do Plenário, 1945 a 1967*. Mimeographed. 2 Vols. Rio de Janeiro, n.d.

Brasil. Ministério do Planejamento e Coordenação Econômica. Escritório de Pesquisa Econômica Aplicada. *Petróleo: Diagnóstico Preliminar. Plano Decenal de Desenvolvimento Econômico e Social*. Brasília, July 1966.

BULHÕES, OCTÁVIO GOUVÊA DE. *À Margem de um Relatório: Texto das Conclusões da Comissão Mista Brasileiro–Americana de Estudos Econômicos (Missão Abbink)*. Rio de Janeiro: Edições Financeiras, 1950.

DA ROCHA, D. FLEURY. "Pesquisas de Petróleo em São Paulo". *Boletim do Ministério da Agricultura*, No. 22 (1933), pp. 463–72.

"Entrevista do General Albino Silva, Presidente da Petrobrás, sôbre o Relatório dos Técnicos Soviéticos em 11-10-1963". Typewritten. Rio de Janeiro, n.d.

FRÓES ABREU, SÍLVIO. *Recursos Minerais do Brazil: Combustíveis Fósseis e Minérios Metálicos*. Vol. II. Rio de Janeiro: Ministério da Indústria e do Comércio, Instituto Nacional de Tecnologia, 1962.

Fundação Getúlio Vargas. *A Missão Cooke no Brasil*. Rio de Janeiro, 1949.

Os Fundamentos da "Petrobrás". Rio de Janeiro: Imprensa Nacional, 1952.

International Petroleum Encyclopedia, 1970. Tulsa: Petroleum Publishing Co., 1969.

OLIVEIRA, AVELINO IGNACIO DE. "Situação do Problema do Petróleo no Brasil em 1938". República dos Estados Unidos do Brasil. Ministerio da Agricultura. Departamento Nacional da Producção Mineral. Serviço de Fomento da Producção Mineral. *Boletim N. 23*. Rio de Janeiro, 1938.

PAIVA, GLYCON DE, and DO AMARAL, IRNACK C. "Rumos Novas em Sondagens Profundas", Brasil. Ministerio da Agricultura. *Boletim da Divisão de Fomento da Producção Mineral, Departamento Nacional da Producção Mineral*, No. 36 (1939).

Petróleo Brasileiro S.A. *Exercício de 1966*. Rio de Janeiro, February 2, 1967.
―――――. *1965: How Was It to Petrobrás*. Rio de Janeiro, n.d.
―――――. *Petróleo: Legislação Básica*. Rio de Janeiro, 1965.
―――――. *Relatório*. Rio de Janeiro, 1954―.
PINTO, MARIO DA SILVA. "Os Óleos do Lobato". Brasil. Ministerio da Agricultura. Departamento Nacional da Producção Mineral. Laboratório Central da Producção Mineral. *Avulso N° 3*. Rio de Janeiro, 1939.
4° Congresso Internacional de Petróleo. *Brasil: O Petróleo da Amazônia*. Rome, 1955.
United States Department of Commerce. Bureau of Foreign and Domestic Commerce. M. A. Cremer. "Petroleum in Brazil". Washington: G.P.O., 1925.
WASHBURNE, CHESTER W., "Petroleum Geology of the State of São Paulo". *Boletim 22 da Commissão Geographica e Geologica do Estado de São Paulo*. São Paulo, 1930.
WHITE, ISRAEL CHARLES. *Final Report (Relatorio Final)*. Rio de Janeiro: Imprensa Nacional, 1908.

Newspapers and Journals

I used the complete file of the *New York Times* as a source, supplemented by the nationalistic newspaper *Última Hora*, 1956―65. On significant issues and events, the newspapers *Jornal do Brasil*, *O Jornal*, and *O Globo* were consulted for other points of view. The more nationalistic arguments were presented by *Jornal de Debates* and *O Semanário*, both of which were consulted on significant issues. The whole was supplemented by the useful clippings collection found in the Information Section, Statistical Council, IBGE, Rio de Janeiro.

The following journals were consulted in their entirety (unless otherwise indicated):

Bulletin of the American Association of Petroleum Geologists, 1949―66.
Conjuntora Econômica. Rio de Janeiro: Fundação Getúlio Vargas, 1947―.
International Petroleum Quarterly. U.S. Bureau of Mines.

Oil and Gas Journal. Tulsa, 1902— .
O'Shaughnessy's Oil Report (O'Shaughnessy's Oil Bulletin after 1931). New York, 1926—30.
Petrobrás. Rio de Janeiro, 1954— .
Petroleum Engineer. Tulsa, 1929— .
Petroleum Press Service.
Revista do Clube de Engenharia. Rio de Janeiro.
Revista do Clube Militar. Rio de Janeiro.
World Oil (Oil Weekly to 1940). Houston, 1916— .
World Petroleum. New York, 1930— .
World Petroleum Report. New York, 1947— .

Other journals were consulted for specific articles, as below.

Articles

BONILLA, FRANK. "A National Ideology for Development". In American Universities Field Staff. *Expectant Peoples*. New York: Vintage, 1967, pp. 232—64.

BRAGA, ODILON. "O Problema do Petróleo". *Mineração e Metalurgia*, Vol. XIII, No. 73 (May—June 1948), pp. 37—42.

"Brazil: the Petrochemicals Industry". *B.O.L.S.A. Review*, Vol. IV (June 1970), pp. 312—17.

BRETON, ALBERT. "The Economics of Nationalism". *Journal of Political Economy*, Vol. LXXII (August 1964), pp. 376—86.

CAMPOS, ROBERTO DE OLIVEIRA. "As Três Falácias do Momento Brasileiro". *Digesto Econômico*, Vol. XIII, No. 133 (January—February 1957), pp. 49—66.

CHANDLER, GEOFFREY. "The Myth of Oil Power". *International Affairs*, Vol. XLVI (1970), pp. 710—18.

DO AMARAL, IRNACK CARVALHO, and PAIVA, GLYCON DE. "Considerações Geofísicas e Geológicas para Pesquisa de Petróleo". *Mineração e Metalurgia*, Vol. II, No. 11 (January—February 1938), pp. 343—51.

FERNANDES, GERSON. "O Problema do Petróleo no Brasil: Situação Atual e Possibilidades Futuras". *Boletim Geográfico*, Vol. XVIII, No. 159 (November—December 1960), pp. 1004—16.

FRÓES ABREU, SÍLVIO. "Aspectos Geográficos Geológicos e Políticos da Questão do Petróleo no Brasil". *Revista Brasileira de Geografia*, Vol. VIII, No. 4 (October—December 1946), pp. 509—34.

_____. "O Problema do Petróleo no Brasil". *Revista Brasileira de Geografia*, Vol. 1, No. 2 (September–October 1938), pp. 53–71.

_____. "O Recôncavo da Bahia e o Petróleo de Lobato". *Boletim Geográfico*, Vol. VI, No. 70 (January 1949), pp. 1148–71.

GUDIN, EUGÊNIO. "Petrobrás: Desperdicio e Incapacidade". *Digesto Econômico*, Vol. XVII, No. 158 (March–April 1961), pp. 99–101.

LIMA, HERMES. "Novos e Velhos Enganos sôbre o Nacionalismo". *Carta Mensal*, Vol. V, No. 54 (September 1959), pp. 33–44.

LINK, WALTER K. "Exploration Brazil 1955". *International Geological Congress, 20th, Petroles y Gas Symposium*, Vol. IV (Mexico 1956), pp. 61–63.

MARQUES, ELÁDIO. "A Realidade sôbre o Petróleo do Brasil". *O Observador Econômico e Financeiro*, Vol. XX, No. 231 (May 1955), pp. 54–60; No. 232 (June 1955), pp. 24–31.

MOURA, PEDRO DE, and FERNANDES, GERSON. "Petroleum Geology in the State of Bahia". *International Geological Congress, 19th, Algeria*. C.R., Sec. 14, F.16 (1953), pp. 65–84.

ODDONE, D. S. "Oil Prospects in the Amazon Region". *International Geological Congress, 19th, Algeria*. C.R., Sec. 14, F.16 (1953), pp. 247–72.

OLIVEIRA, EUZEBIO DE. "Pesquisas de Petróleo". *Anais da Escola de Minas de Ouro Prêto*, Vol. XV (1917), pp. 105–16.

PAIVA, GLYCON DE. "A Descoberta de Petróleo no Brasil e Suas Imediatas Consequências". *Mineração e Metalurgia*, Vol. III, No. 18 (March–April 1939), pp. 335–44.

_____. "Petrobrás como Banco da Subversão Nacional e Escola Prática de Corrupção". *Digesto Econômico*, Vol. XX, No. 176 (March–April 1964), pp. 63–68.

_____. "O Petróleo de Nova Olinda". *Digesto Econômico*, Vol. XI, No. 125 (September–October 1955), pp. 138–42.

_____. "O Petróleo de Nova Olinda". *Carta Mensal*, Vol. I, No. 3 (June 1955), pp. 36–44.

_____. "Princípios sôbre a Geologia do Petróleo". *Carta Mensal*, Vol. V, No. 49 (January–April 1959), pp. 23–40.

_____. "Regime Legal e Produção de Petróleo". *Revista Brasileira de Economia*, Vol. VI, No. 2 (June 1952), pp. 7–67.

"Política do Petróleo". *Cadernos do Nosso Tempo*, No. 4 (April–August 1955), pp. 35–56.

SAMPAIO, ALDE. "Nacionalismo Econômico". *Digesto Econômico*, Vol. XIX (January–February 1962), pp. 31–45.

SCHIFFINO, RINALDO. "Problemas do Comércio e da Industrialização do Petróleo no Brasil". *Mensagem Econômica*, Vol. VII, No. 68 (August 1958), pp. 13–20.

SHERMAN, W. B. "Petroleum Possibilities of the Gondwana System of Brazil". *International Geological Congress, 19th, Algeria*, C.R., Sec. 14, F.16 (1953), pp. 85–96.

"Situação Econômica e Política de Desenvolvimento". *Cadernos do Nosso Tempo*, No. 5 (1956), pp. 67–119.

SMITH, PETER SEABORN. "Bolivian Oil and Brazilian Economic Nationalism". *Journal of Inter-American Studies and World Affairs*, Vol. XIII (1971), pp. 166–81.

———. "Petrobrás: the Politicizing of a State Company, 1953–1964". *The Business History Review*, Vol. XLVI (1972), pp. 182–201.

TÁVORA, JUAREZ. "A Batalha do Petróleo; Notas a Margem do Tema 'Quem Matou Vargas' ". *Manchete*, No. 822–826 (January 20–February 17, 1968).

TORRES, LÚCIO GLAUCO. "Petróleo da Bolivia". *O Observador Econômico e Financeiro*, No. 279 (May 1959), p. 52.

TOURINHO, BORBA. "Petróleo no Brasil". *O Observador Econômico e Financeiro*, Vol. XXI, No. 246 (August 1956), pp. 42–47; No. 247 (September 1956), pp. 72–79.

Personal Interviews

Barroso, Eng. Geonísio Carvalho. Rio de Janeiro, July 7, 1970.
Cordeiro, Oscar. Salvador, Bahia, April 11, 1967.
do Amaral, Dr. Irnack Carvalho. Rio de Janeiro, October 31, 1967; June 29 and July 3 and 6, 1970.
Link, Walter K. LaPorte, Indiana, September 16, 1966.
Monteiro, Dr. Lauriston Pessôa. Rio de Janeiro, November 6, 1967; July 15, 1970.
Paiva Teixeira, Dr. Glycon de. Rio de Janeiro, November 3 and 10, 1967; July 6, 1970.
Távora, Gral. Juarez. Rio de Janeiro, July 8, 1970.

Pamphlets and Collected Documents

ALVES, SENADOR LANDULPHO. *O Problema Brasileiro do Petróleo*. Rio de Janeiro: Livraria José Olympio Editôra, 1954.

BARROSO, GEONÍSIO CARVALHO. *Ação da Petrobrás no Recôncavo Bahiano.* São Paulo: Serviço de Publicações do Centro e Federação das Indústrias do Estado de São Paulo, 1958.

———. *O Petróleo no Brasil e Seus Problemas.* Rio de Janeiro: Editôra, Civilização Brasileira S.A., 1957.

BELTRÃO, HÉLIO. *Os Seis Equívocos Fundamentais sôbre a Petrobrás.* Rio de Janeiro, 1957.

BERNARDES, ARTUR. *Batalha do Petróleo.* Rio de Janeiro, 1948.

BRAGA, VALÉRIO. *Conferências de Estudo do Problema do Petróleo.* Rio de Janeiro, 1952.

CHAMMA, JORGE ABDALLA. *Por um Brasil Melhor.* Rio de Janeiro: Livraria Clássica Brasileira, 1955.

Comité de Trabalhadores pró Emancipação Econômica do Brasil. *Petróleo para o Povo Brasileiro.* Rio de Janeiro, 1952.

DO NASCIMENTO, MAJOR ITURIEL. *O Brasil e o Seu Petróleo.* Rio de Janeiro: Livraria Clássica Brasileira, 1948.

DUQUE-ESTRADA, RODRIGO. *O Petróleo no Brasil: Holding de Estado.* Rio de Janeiro: Livraria-Editôra da Casa do Estudante do Brasil, 1949.

FIGUEIREDO, GRAL. MÁRIO POPPE DE. *Combustíveis Líquidos.* Rio de Janeiro: Imprensa Nacional, 1958.

FONTES, LOURIVAL. *Política, Petróleo e População.* Rio de Janeiro: Livraria José Olympio Editôra, 1958.

FREITAS, GRAL. TÁCITO. *Petróleo apesar de Mr. Link.* Rio de Janeiro: Edições Gernasa, 1964.

GREENHALGH, JUVENAL. *O Problema do Petróleo Nacional.* Rio de Janeiro, 1948.

GUDIN, EUGÊNIO. *Análise de Problemas Brasileiros, 1958 – 1964.* Rio de Janeiro: Agir, 1965.

GUILHERME, OLYMPIO. *A Verdade sôbre Roboré.* Rio de Janeiro: Livraria Freitas Bastos, 1960.

HENNING, ANTÔNIO JOSÉ. *Influência dos Trusts na Indústria Petrolífera.* Rio de Janeiro, 1952.

HORTA BARBOSA, GRAL. J. C. *Nuestro Petróleo Debe Ser Explotado por el Gobierno.* Buenos Aires, 1948.

———. *Problemas do Petróleo no Brasil.* Rio de Janeiro, 1947.

Instituto Brasileiro de Petróleo. *Economia do Petróleo.* Rio de Janeiro, 1959.

LACERDA, HÉLIO DE. *Petróleo e Outros Problemas: Democracia, Reforma Agrária, Industrialização. Artigos, Discursos, Conferências e Entrevistas, 1947 – 1948.* São Paulo, n.d.

LEITE, SOLIDÔNIO. *O Petróleo e o Dever do Brasil.* Rio de Janeiro: J. Leite & Cia., 1927.

LUDOLF, MÁRIO LEÃO. *Da Conveniência de Exploração do Petróleo Boliviano.* Rio de Janeiro: Confederação Nacional da Indústria, Conselho Econômico, 1959.

MELLO E SILVA, DRAULT ERNANNY DE. *A Questão do Petróleo.* Rio de Janeiro, 1952.

MIRANDA, JACY VIEIRA DE. *Planejamento da Indústria do Petróleo.* Rio de Janeiro, November 25, 1966.

MOURA, PEDRO DE. *Aspectos Técnicos do Problema do Petróleo no Brasil.* Rio de Janeiro, 1948.

NERY, ADALGISA. *Retrato sem Retoque.* Rio de Janeiro: Editôra Civilização Brasileira, 1963.

NUNES, JANARY GENTIL. *Defesa dos Programas da Petrobrás.* Rio de Janeiro, 1959.

PACHECO, J. JANOT. *Diversos Aspectos do Problema do Petróleo: Uma Solução Nacional.* Rio de Janeiro, 1947.

PASSOS, GABRIEL DE REZENDE. *Estudo sôbre o Acordo de Roboré.* São Paulo: Editôra Fulgor, 1960.

———. *Minerais Atômicos.* Rio de Janeiro: Livraria José Olympio Editôra, 1959.

———. *Temas Nacionalistas.* São Paulo: Editôra Fulgor, 1961.

Petróleo Brasileiro S.A. *Esclarecimentos Prestados a Comissão Parlamentar de Inquérito.* Rio de Janeiro, 1958.

POGUE, JOSEPH E. *Oil in Brazil.* New York, 1951.

[Refinaria e Exploração de Petróleo União]. *Petróleo Boliviano e Mercado Brasileiro.* São Paulo, 1956.

ROMARIZ, JOÃO DE. *Palestra Realizada na Rádio Glôbo sôbre o Problema do Nosso Petróleo.* Rio de Janeiro, May 21, 1948.

SIMÕES LOPES, ILDEFONSO. *O Petróleo Brasileiro.* Rio de Janeiro: Imprensa Nacional, 1945.

TÁVORA, SENADOR FERNANDES. *Como Poderemos Resolver o Problema do Petróleo no Brasil.* Rio de Janeiro, 1949.

TÁVORA, JUAREZ. *O Petróleo do Brasil.* São Paulo: Fulgor, 1947.

———. *Petróleo para o Brasil.* Rio de Janeiro: Livraria José Olympio Editôra, 1955.

———. *O Problema Brasileiro do Petróleo: Ensaio de Solução Objetiva.* Rio de Janeiro: Livraria José Olympio Editôra, 1948.

TUPINAMBÁ, TARCÍSIO. *Petróleo e Liberdade.* Niteroi, 1950.

TUPINIQUINS, TAPIRAPÉ. *Petróleo: Sinônimo de Lutas*. Niteroi, 1957.
VARGAS, GETÚLIO. *A Campanha Presidencial*. Rio de Janeiro: Livraria José Olympio Editôra, 1951.
———. *O Govêrno Trabalhista no Brasil*. 2 Vols. Rio de Janeiro: Livraria José Olympio Editôra, 1952 and 1954.
———. *A Nova Política do Brasil*. Vol. VIII. Rio de Janeiro: Livraria José Olympio Editôra, 1941.
———. *A Política Nacionalista do Petróleo no Brasil*. Rio de Janeiro: Livraria José Olympio Editôra, 1964.

Biographies and Memoirs

CAFÉ FILHO, JOÃO. *Do Sindicato ao Catete: Memórias Políticas e Confissões Humanas*. 2 Vols. Rio de Janeiro: Livraria José Olympio Editôra, 1966.
CASTRO E SELIVA, E. M. DE. *Nossos Problemas do Petróleo*. Rio de Janeiro, Editôra Melso, 1964.
CAVALHEIRO, EDGARD. *Monteiro Lobato: Vida e Obra*. 2 Vols. São Paulo: Edição da Companhia Distribuidora de Livros, especialmente para a Companhia Editôra Nacional, 1955.
HENRIQUES, AFFONSO. *Ascensão e Queda de Getúlio Vargas*. Vol. III. Rio de Janeiro: Distribuidora Record, 1966.
MAGALHÃES, JURACI. *Minha Vida Publica na Bahia*. Rio de Janeiro: Livraria José Olympio Editôra, 1957.
SODRÉ, NELSON WERNECK. *Memórias de um Soldado*. Rio de Janeiro: Civilização Brasileira, 1967.
VEGARA, LUIZ. *Fui Secretário de Getúlio Vargas: Memórias dos Anos de 1926–1954*. Rio de Janeiro: Livraria José Olympio Editôra, 1960.

General Works

ALEXANDER, ROBERT J. *The Bolivian National Revolution*. New Brunswick, N.J.: Rutgers University Press, 1958.
ALMEIDA, A. ALVES DE. *Petróleo em Terras do Brasil*. São Paulo, 1948.
ALMEIDA, CÂNDIDO ANTÔNIO MENDES DE. *Nacionalismo e Desenvolvimento*. Rio de Janeiro: Instituto Brasileiro de Estudos Afro–Asiáticos, 1963.

ARAUJO CASTRO, RAIMUNDO DE. *A Nova Constituição Brasileira*. Rio de Janeiro, 1935.

BAER, WERNER. *The Development of the Brazilian Steel Industry*. Nashville: Vanderbilt University Press, 1969.

———. *Industrialization and Economic Development in Brazil*. New Haven: Yale University Press, 1965.

BAKLANOFF, ERIC N. ed. *New Perspectives of Brazil*. Nashville: Vanderbilt University Press, 1966.

BELLO, JOSÉ MARIA. *A History of Modern Brazil, 1889–1964*. Stanford: Stanford University Press, 1966.

BURNS, E. BRADFORD. *Nationalism in Brazil: A Historical Survey*. New York: Frederick A. Praeger, 1968.

CALÓGERAS, JOÃO PANDIÁ. *A History of Brazil*. New York: Russell & Russell, 1963.

CARVALHO, EDSON DE. *O Drama da Descoberta do Petróleo Brasileiro*. São Paulo: Editôra Brasilense, 1954.

CARVALHO, ESTÉVÃO LEITÃO DE. *Petróleo! Salvação ou Desgraça do Brasil?* Rio de Janeiro: Edição do Centro de Estudos e Defesa do Petróleo e da Economia Nacional, 1950.

COHN, GABRIEL. *Petróleo e Nacionalismo*. São Paulo, 1968.

COUTINHO, LOURIVAL, and SILVEIRA, JOEL. *História de uma Conspiração (Bolívia, Brasil e Petróleo)*. Rio de Janeiro: Editôra Coelho Branco, 1959.

———. *O Petróleo do Brasil: Traição e Vitória*. Rio de Janeiro: Editôra Coelho Branco, 1957.

DA FONSECA, GONDIN. *Que Sabe Você sôbre Petróleo?* 3rd ed. Rio de Janeiro: Livraria São José, 1955.

DALAND, ROBERT T. *Brazilian Planning: Development Politics and Administration*. Chapel Hill: University of North Carolina Press, 1967.

DALEMONT, ETIENNE. *O Petróleo*. São Paulo: Editôra Fulgor, 1961.

DECHERT, CHARLES R. *Ente Nazionale Idrocarburi: Profile of a State Corporation*. Leiden, 1963.

DUBNIC, VLADIMIR REISKY DE. *Political Trends in Brazil*. Washington: Public Affairs Press, 1968.

DULLES, JOHN W. F. *Unrest in Brasil: Political–Military Crises 1955–1964*. Austin: University of Texas Press, 1970.

———. *Vargas of Brazil: A Political Biography.* Austin: University of Texas Press, 1967.

FANNING, LEONARD M. *American Oil Operations Abroad.* New York: McGraw-Hill, 1947.

FEIS, HERBERT. *Petroleum and American Foreign Policy.* Stanford: Stanford University Press, 1944.

FRANKEL, P. H. *The Essentials of Petroleum: A Key to Oil Economics.* London: Cass, 1969.

FRÓES ABREU, SÍLVIO, PAIVA, GLYCON DE, and DO AMARAL, IRNACK CARVALHO. *Contribuições para a Geologia do Petróleo no Recôncavo, Baía.* Rio de Janeiro: Oficinas Gráficas do Serviço de Publicidade Agrícola, 1936.

GONSALVES, ALPHEU DINIZ. *O Petróleo no Brasil.* Rio de Janeiro: Livraria José Olympio Editôra, 1963.

HARTSHORN, J. E. *Oil Companies and Governments: An Account of the International Oil Industry in Its Political Environment.* London: Faber and Faber, 1962.

HIRST, DAVID. *Oil and Public Opinion in the Middle East.* London: Faber and Faber, 1966.

HOROWITZ, IRVING L. *Revolution in Brazil: Politics and Society in a Developing Nation.* New York: E. P. Dutton & Co., 1964.

IRELAND, GORDON. *Boundaries, Possessions, and Conflicts in South America.* Cambridge: Harvard University Press, 1938.

JAGUARIBE, HÉLIO. *O Nacionalismo na Atualidade Brasileira.* Rio de Janeiro: Instituto Superior de Estudos Brasileiros, 1958.

JAMES, PRESTON. *Latin America.* 3rd ed. New York: Odyssey Press, 1959.

JOHNSON, HARRY G., ed. *Economic Nationalism in Old and New States.* Chicago: University of Chicago Press, 1967.

KEITH, HENRY H., and EDWARDS, S. F., eds. *Conflict and Continuity in Brazilian Society.* Columbia: University of South Carolina Press, 1969.

LEFF, NATHANIEL H. *Economic Policy-Making and Development in Brazil, 1947–1964.* New York: John Wiley & Sons, 1968.

LENCZOWSKI, GEORGE. *Oil and State in the Middle East.* Ithaca: Cornell University Press, 1960.

LEVY, W. J., INC. *The Search for Oil in Developing Countries: A Problem* of Scarce Resources and Its Implications for State and Private Enterprise. Prepared at the request of the International Bank for Reconstruction and Development. New York, 1960.

LIMA, ILDEU RAMOS DE. *Petróleo para o Brasil: Industrialização das Rochas Pirobetuminosas.* Rio de Janeiro, 1951.

LIMA, MEDEIROS. *Petróleo: Desenvolvimento ou Vassalagem? (A Defecção de Frondizi).* Rio de Janeiro: Antunes, 1960.

LINS, ALVARO. *Rio-Branco.* 2 Vols. Rio de Janeiro: Livraria José Olympio Editôra, 1945.

LONGRIGG, STEPHEN H. *Oil in the Middle East: Its Discovery and Development.* London: Oxford University Press, 1968.

MANGABEIRA, FRANCISCO. *Imperialismo, Petróleo, Petrobrás.* Rio de Janeiro, 1964.

MARINHO, ILMAR PENNA, JR. *Petróleo — Soberania e Desenvolvimento.* Rio de Janeiro: Edições Bloch, 1970.

MASCARENHAS, ANDERSON O. *Roboré: um Torpedo contra a Petrobrás.* São Paulo: Editôra Fulgor, 1959.

MASON, EDWARD S. *Economic Planning in Underdeveloped Areas: Government and Business.* New York: Fordham University Press, 1958.

MAYA, EMILIO DE. *O Brasil e o Drama do Petróleo.* Rio de Janeiro: Livraria José Olympio Editôra, 1938.

MONTEIRO LOBATO, JOSÉ BENTO. *O Escândalo do Petróleo e Ferro.* São Paulo: Companhia Editôra Nacional, 1936.

NEEDLER, MARTIN C. *Latin American Politics in Perspective.* Princeton: D. Van Nostrand Co., 1963.

O'CONNOR, HARVEY. *World Crisis in Oil.* New York: Monthly Review Press, 1962.

ODELL, PETER R. *An Economic Geography of Oil.* London: G. Bell, 1963.

———. *Oil and World Power: A Geographical Interpretation.* Harmondsworth: Penguin Books, 1970.

OLIVEIRA, EUZEBIO PAULO DE. *História da Pesquisa de Petróleo no Brasil.* Rio de Janeiro: Oficinas Gráficas do Serviço de Publicidade Agrícola, 1940.

PENROSE, EDITH T. *The Large International Firm in Developing Countries: The International Petroleum Industry.* London: Allen & Unwin, 1968.

PEREIRA, OSNY DUARTE. *Estudos Nacionalistas: As Cadeias do Imperialismo no Brasil.* 2 Vols. São Paulo: Editôra Fulgor, 1960.
PIMENTEL, J. F. DE BARROS. *O Problema do Petróleo no Brasil.* Rio de Janeiro: Imprensa Nacional, 1948.
POPPINO, ROLLIE E. *International Communism in Latin America: A History of the Movement.* New York: Free Press, 1964.
RABINOVITCH, MOISES. *Pró Solução Propria do Problema do Petróleo Economico Nacional (Localização do Oleo Fossil no País).* Rio de Janeiro, 1937–38.
SCHNEIDER, RONALD M. *The Political System of Brazil: Emergence of a "Modernizing" Authoritarian Regime, 1964–1970.* New York: Columbia University Press, 1971.
SELL, GEORGE. *The Petroleum Industry.* London: Oxford University Press, 1963.
SKIDMORE, THOMAS E. *Politics in Brazil, 1930–1964: An Experiment in Democracy.* New York: Oxford University Press, 1967.
SMITH, PETER SEABORN. "Petroleum in Brazil: A Study in Economic Nationalism". Ph.D. dissertation, University of New Mexico, 1969.
SODRÉ, NELSON WERNECK. *História Militar do Brasil.* Rio de Janeiro: Editôra Civilização Brasileira, 1965.
STEPAN, ALFRED. *The Military in Politics: Changing Patterns in Brazil.* Princeton: Princeton University Press, 1970.
STOCKING, GEORGE WARD. *Middle East Oil: A Study in Political and Economic Controversy.* Nashville: Vanderbilt University Press, 1970.
TANZER, MICHAEL. *The Political Economy of International Oil and the Underdeveloped Countries.* Boston: Beacon Press, 1969.
TENDLER, JUDITH. *Electric Power in Brazil: Entrepreneurship in the Public Sector.* Cambridge: Harvard University Press, 1968.
VAITSMAN, MAURÍCIO. *O Petróleo no Império e na República.* Rio de Janeiro: Seção de Livros da Emprêsa Gráfica O Cruzeiro, 1948.
VICTOR, MÁRIO. *A Batalha do Petróleo Brasileiro.* Rio de Janeiro: Civilização Brasileira, 1970.
VIVACQUA, ATÍLIO. *A Nova Política do Sub-solo e o Regime Legal das Minas.* 2 Vols. Rio de Janeiro: Editôra Panamericana, 1942.
WEISBORD, ALBERT. *Latin American Actuality.* New York: Citadel Press, 1964.

WILSON, CHARLES MORROW. *Oil across the World: The American Saga of Pipelines*. New York: Longmans, Green & Co., 1946.
WIRTH, JOHN D. "Brazilian Economic Nationalism: Trade and Steel under Vargas". Ph.D. dissertation, Stanford University, 1966.
_____. *The Politics of Brazilian Development, 1930–1954*. Stanford: Stanford University Press, 1970.
ZONDAG, CORNELIUS H. *The Bolivian Economy, 1952–65: The Revolution and Its Aftermath*. New York: Frederick A. Praeger, 1966.

INDEX

Acre, 16, 25, 40, 99
Agriculture: Minister of, 9, 18, 22, 23, 29, 30, 34, 46, 52, 53; Ministry of, 10, 11, 14, 18, 22, 27, 30, 34, 38; Secretary of, 15
Agriculture Committee, 18, 19
Alagoas, 9, 11, 23, 24, 25, 30, 35, 40, 99, 113, 123, 175, 176
alcohol, anhydrous, 43, 45
Alliance for Progress, 144
Almeida, Hélio de, 162, 163
Alves, Landulpho, 92, 94
Alves, General (later Marshal) Osvino, 150, 158, 161, 162, 163, 171
Amaral, Irnack Carvalho do, 23, 31, 34-5, 36, 173; Petrobrás director, 100, 116; Petrobrás president, 138, 176
Amazon: River, 71, 92, 99, 114; Valley, 11, 106; Basin, 3, 17, 40, 92, 106, 126, 129, 159; Delta, 177
Amazônas, 17
American and Foreign Power Company (AMFORP), 155
ANCAP, 38
Andrade, Alfredo de, 162
anti-Americanism, 66, 73, 76
anticline (dome), 2, 8
appropriations, 12-13, 40, 47, 56, 69, 72, 96, 99, 166
Aracaju, Sergipe, 42
Aranha, Oswaldo, 96
Araquá, São Paulo, 25, 26, 27, 28, 29, 35
Aratu, Bahia, 42, 48, 99
Argentina, 25, 33, 38, 39, 50, 54, 57, 122, 145

army (armed forces), 27, 34, 41, 44, 49, 55, 72, 89, 101, 104, 106, 109, 111, 118, 135, 137, 147, 150, 152, 164, 170, 171, 172, 186; First Army, 150, 157, 161, 162. *See also* military
Aruba, 42, 123
asphalt, 99, 115, 175
Assessoria Econômica da Presidência da República, 78, 79
Atlantic (oil province), 3, 40
Atlantic Charter, 53
Atlantic Refining Company, 21, 178
Atlas Development, 160

Bahia, 7, 10, 16, 22, 31, 36, 38, 39, 40, 42, 47, 48, 63, 70, 91, 92, 99, 113, 123, 134, 136, 137, 138, 139, 142, 143, 147, 152, 158, 159, 171, 175
Bahia-São Francisco Railway, 7
balance of payments, 78, 91, 163. *See also* foreign debt
Bank of Brazil, 72, 172
Barcellos, Peracchi, 153
Barreirinhas, Maranhão, 3, 159, 176
Barreto, Colonel João Carlos, 46, 47, 51, 52, 53, 79
Barreto, Manhães, 83
Barroso, Geonísio Carvalho, 134, 135, 137, 138, 139
Bastos, General Justino Alves, 121, 141
Bastos, Manoel Ignácio, 22, 29, 31
Belém, Pará, 71
Belo Horizonte, Minas Gerais, 116
Beltrão, Hélio, 117, 124
Benavides, Colonel, 86
Berle, Adolph A., 51

277

Bernardes, Artur, 59, 63, 66, 114; Brazilian President, 15, 114; Senator, 88
Betím, Minas Gerais, 178
Bittencourt, Colonel Alexínio, 117
Bofete, São Paulo, 9
Bolivia, 16, 106, 107, 145; *1938* treaty with Brazil, 34, 107, 118, 119; attempts to implement *1938* treaty (1956–58), 107, 108, 109, 110, 118-21, 182. *See also* Roboré
Bolsa de Mercadorias, 22
Borges, Ortiz, 154
bottlenecks, 75, 111
Braga, Odilon, 30, 52, 53, 55, 56, 61, 62, 65, 66, 69, 83-4
Brantly, J. E., 36
Brasil, General Assis, 22
Brasília, Federal District, 110, 148, 151, 154, 164
Brazilian Academy of Sciences, 31, 37
Brazilian Empire, 8-9
Brazilian Expeditionary Force, 58, 76
Brazilian Press Institute, 135
British Petroleum, 145, 160
Brizola, Leonel, 153, 155-6, 162
business, 105; traditional, export-oriented, 13, 66, 82, 93-4, 98, 112, 168 (*see also* liberalism); newer industrial, 82, 93, 98, 168, 169, 185

Cadernos do Nosso Tempo, 59
Café Filho, João: Vice-President, 96; President, 103, 104, 106, 107, 109, 110, 111
Caixa Econômica Federal, 139
California Transport Corporation, 98
Calógeras, João Pandiá, 9-11, 14, 21, 27
Caloric Company, 21
Camargo, Eugênio Ferreira, 9
Camaú, Bahia, 7
Cameron & Jones Company, 174
Camirí, Bolivia, 107
Campos (sedimentary basin), 3
Campos, Gonzago de, 11-12
Campos, Roberto de Oliveira, 119, 123-4, 125

Campos Sales, Manuel Ferraz de, 9
Candeias, Bahia, 42, 48, 52, 85
Canoas, Rio Grande do Sul, 178
Cantanhede, Plínio Reis de, 79, 106, 108
capital, 1, 11, 19, 48, 83, 109, 129-30, 186; foreign, 18, 19, 33, 40-1, 48, 49, 51, 53, 54, 63, 64, 65, 74, 77, 78, 80, 89, 90, 94, 96, 104, 105, 107, 114, 124, 125, 185; private, 11, 13, 14, 19, 23-4, 25, 26, 27, 28, 29, 31, 33, 36, 48, 49, 51, 52, 54, 56, 60, 62, 63, 65, 71, 78, 80, 81, 82, 84, 89, 92, 100, 118-21, 122, 124, 144, 157, 185; "mixed", 44, 48-9, 51, 54, 55, 81, 124; government, 56, 80, 81, 124, 165. *See also* foreign investment
"Captain X", 65
Capuava, São Paulo, 51, 71, 114, 128, 141, 155, 156, 157, 171, 173
Cardoso, General Felicíssimo, 82
Cardoso, Graccho, 19
Carmópolis, Sergipe, 175
Carvalho, Edson de, 23, 24, 25, 30
Castello Branco, General Humberto de, 171, 173, 174, 176
Catete Palace, 20
Catholic Left, 140
Cavalcanti, Kerginaldo, 93, 94
Ceará, 15, 86, 92
Central Laboratory of Mineral Production, 22
Centro de Estudos e Defensa do Petróleo e da Economia Nacional. See Petroleum Centre
Chamber of Deputies, 18, 19, 52, 55, 61, 62, 65, 66, 67, 79, 82, 84, 86, 87, 88, 89, 90, 91, 92, 93, 94, 95, 96, 98, 108, 111, 120, 125, 126, 134, 148, 149, 150, 151, 171
Chamma, Jorge Abdalla, 64
Chapultepec Conference, 53
Chase Manhattan Bank, 145
Chateaubriand, Assis, 92, 93, 105, 115
chauvinism, 12, 13, 73, 120, 164, 185, 186. *See also* nativism, xenophobia
Cinelândia plaza (Rio de Janeiro), 113

Clube Militar, 64-5, 72, 85, 86, 141, 153; debates over oil policy, 64; *1950* election, 76; *1958* election, 112, 121, 141; *1962* election, 141
coal, 7, 10, 11, 13, 69, 90
coffee, 13, 15, 20, 78, 91, 117, 131
Colombia, 62
colonial(ism), 58, 59, 60, 73, 167, 169. *See also* neocolonialism
Comissão de Anteprojeto da Legislação do Petróleo, 52, 53, 55, 56
Committee of Foreign Commerce, 174
communism, 55, 56, 65, 66, 73, 76, 78, 85, 87, 89, 90, 93, 105, 121, 150, 153, 156, 163, 173
Communist Party, Brazil, 56, 121
Companhia Brasileira de Petróleo, 32
Companhia de Petróleo da Amazônia, 114
Companhia Geral de Petróleo Pan-Brasileira, 32
Companhia Petróleo Nacional, 23-4, 30, 35
Companhia Petróleos do Brasil, 25, 26, 27, 28, 29, 35
"completion of the state monopoly", 108, 146, 154, 157, 164, 174
concession for petroleum exploitation, 53, 54, 55, 61-2, 63, 66, 77, 81, 87, 92, 95, 107, 109, 122
Congress, 20, 69, 70, 71, 72, 73, 78, 82, 84, 89, 91, 92, 93, 120, 125, 127, 135, 137, 145, 148, 150, 186. *See also* Chamber of Deputies, Senate
Constituent Assembly, 50
Constitution: *1891*, 8, 9, 13, 14, 19, 21; *1934*, 27, 33; *1937* (Estado Nôvo), 33, 34, 41, 43, 46; *1946*, 49, 52, 111
Constitution, U.S.S., 146
Consultoria Econômica (Petrobrás), 148
continental shelf, 175, 176, 182, 188
contract, 41, 42, 87, 95, 98, 99, 123, 145, 160-1, 163, 173, 174, 182
Contribuições para a Geologia do Petróleo no Recôncavo, Baía, 31, 37

Convention of Petroleum Defence, 86, 89
Cordeiro, Oscar, 23-4, 29, 30, 31, 36
corporatism, 21, 27, 33, 38, 39, 44, 49, 58, 73, 81, 183, 185, 186
Corrêa, Antônio, 86
Correio da Manhã, 63, 66, 104-5
Correio do Povo, 154
corruption, 143, 161
Costa, General Canrobert Pereira da, 100, 106, 109
Costa, Fernando, 34
Costa, General Zenóbio da, 85
Council of Taxpayers, 21
coup, 152, 161; *1930*, 21, 186 (*see also* Revolution of 1930); *1937*, 33, 34, 186; attempted *1955*, 111; *1964*, 165, 171, 172, 176, 177, 178, 186
cracking, 47
Creole Petroleum Corporation, 123
crude oil, 33, 38, 41, 47, 48, 50, 51, 54, 55, 67, 68, 70, 79, 98, 102, 114, 116, 118, 124, 128, 129, 131, 134, 135, 143, 145, 157-8, 163, 172, 173, 174, 179, 181, 183, 184, 188; exchange of Brazilian for foreign, 116, 123, 135, 188-9
Cubatão, São Paulo, 79, 98, 102, 103, 114, 142, 147, 156, 158, 163, 178
Curtice, Arthur A., 62-3
Czechoslovakia, 71

Dantas, San Tiago, 15
decree, presidential, 21, 34, 38-9, 164, 171-2, 173, 174
Decree-Laws: *366*, 35, 36; *395*, 35, 39, 41; *538*, 35; *3701*, 37
defence, national, 18, 19, 52, 53, 55, 61, 74, 109. *See also* security, national
DeGolyer, Everett L., 47
DeGolyer and MacNaughton, 47, 68
democracy, 60, 169
DEPEX (Petrobrás's Exploration Department), 102, 125, 126, 127, 128, 131, 136, 160
Depression, 20
DEPRO (Petrobrás's Production Department), 160

Derby, Orville, 10
derivatives (products), 41, 77, 78, 80, 83, 108, 116, 129-30, 134, 139, 150, 154, 164, 173, 177-8, 180
developmentalist nationalism, 59, 73, 76, 111, 112, 113, 120, 125, 127, 151, 164, 185, 186
Devonian period, 15
diabase (crystalline rock), 8, 24, 28, 29
Diário de Notícias, 59, 63, 105, 115-16
distillation, 8, 33, 144
distillery, 33, 48, 174, 177
distribution of petroleum derivatives, 21, 38, 56, 67, 79, 81, 82, 83, 87, 90, 108, 131, 146, 150-1, 154, 155, 178, 181; nationalization, 128, 146, 150, 164, 171-2, 178-9. *See also* Petrobrás
Dom João, Bahia, 51
drilling, 11, 13, 24, 26, 36-7, 38, 40, 41, 63, 68, 69, 72, 91, 99, 106, 115, 116, 121, 126, 127, 142, 143, 149, 158, 159, 160, 176, 177
Drilling and Exploration Company, 36, 40, 42, 47
Dulles, John Foster, 122
Duque de Caxias, Rio de Janeiro, 114, 135, 141, 142, 147, 148, 158, 163, 170, 174, 178
Dutra, General Eurico Gaspar: Minister of War, 44; President, 49, 52, 70, 72, 73, 74, 99, 168

Economic Commission for Latin America, 59
economic emancipation, 80, 83, 101, 108, 111, 113, 118, 150, 151, 168
economic nationalism, 46, 53, 54, 55, 59, 66, 72, 76, 107, 168, 169
education, 178
Electric Power in Brazil, 168
electricity, 78, 99, 142, 168-9. *See also* hydro-electric power
Emancipação, 86
Employees' Association of Petrobrás, 153-4
Empresa Paulista de Petróleos, 11
energy, 10, 13, 60, 70, 75, 77, 78, 80, 111, 168, 183, 184, 186

Engineering Club, 53
Engineers' Association of Rio de Janeiro and Guanabara, 156
England, 8, 61. *See also* Great Britain
Ente Nazionale Idrocarburi (ENI), 110, 115, 166
entreguista, 89, 91, 92, 104, 117, 119, 122, 134, 139, 142, 143, 150, 153, 164
equipment, 38, 42, 47, 51, 54, 68, 69, 71, 91, 97, 98, 99, 115, 122, 128, 131, 149, 159, 163, 173; CNP and Petrobrás purchases in Brazil, 91, 142, 178
Espírito Santo: sedimentary basin, 3; state, 177
Esso Export Corporation, 98, 123
O Estado de São Paulo, 28, 93
Estado Nôvo, 33, 34, 38, 49, 52, 53, 66, 76, 78
Etchegoyen, General Alcides, 86
Europe, 51, 62, 115, 119, 122, 188
Exploration Service of the Amazon, 106
Export-Import Bank, 51
exports, 13, 20, 54, 91, 109, 118, 135, 178
expropriation. *See* nationalization

Faraco, Daniel, 95
Farias, Jairo José, 152, 160-2, 171
FAROS, Sociedad Anónima, 145
fascism, 49, 58, 105
Federal District, 89, 90, 137
Federation of Commercial Associations, 104, 181
Federation of Industries, 140
fertilizer, 79
Figueiredo, João Neiva de, 100, 115-16
Figueiredo, General Mário Poppe de, 130
figurehead (*testa de ferro*), 26, 32, 72, 79, 89, 119, 160
Finance, Minister of, 96, 104, 105, 107, 155
First World War, 10
Fleiuss, General Henrique, 117
Fleury da Rocha, Domingos, 23, 28, 29

Index 281

Folha da Manhã, 90
Fonseca, Osvaldo, 88
Fontoura, João Neves da, 78
foreign debt, 91, 120, 163, 173. See also balance of payments
foreign exchange (hard currency, dollars), 52, 70, 71, 72, 77, 78, 79, 91, 97, 98, 103, 109, 110, 117, 118, 119, 120, 121, 123, 124, 131, 134, 145, 151, 158, 160, 163, 168, 172, 179, 184
foreign investment, 5, 19, 37, 41, 46, 50, 54, 60, 61, 62, 72, 77, 78, 79, 81, 83, 86, 88, 90, 96
Foreign Office. See Itamarati
foreigners, 12, 13, 14, 17, 18, 19, 21, 28, 34, 47, 61, 62, 66, 78, 88, 94, 95, 99, 100, 103, 110, 114, 117, 121, 166, 169, 173, 181, 184, 185
Forrestal, James, 57
Fortune, 179
Foster-Wheeler Corporation, 71, 114
France, 61, 71, 158
Freitas, General Tácito, 128, 159, 160
Fróes Abreu, Sílvio, 30, 31, 32, 36, 37, 49
FRONAPE (National Tanker Fleet), 160-1, 179
Frondizi, Arturo, 122
frontiers, 12, 16, 18, 19
fuel oil, 33, 42, 45, 48, 172
fuel shortage, 42-3, 44, 53, 78

Galdo, Colonel Lívio, 173
Ganso Azul, 92, 114
gasogênio, 43
gasoline, 33, 42, 43, 45, 47, 170, 172
Geisel, Ernesto, 128
General Directory of Mineral Production (DGPM), 23
General Labour Command (CGT), 157
General Staff (Army), 63, 94
Geological and Mineralogical Service of Brazil (SGM), 10, 11, 12, 13, 14, 16, 17, 18, 22, 23, 24, 25, 26, 34
Geological Service of São Paulo, 11
geology, petroleum, 2, 69, 116; characteristics of Brazilian, 3, 14, 51, 68, 160, 177, 183; Brazilian knowledge about, awareness of, 3-4, 11, 19, 32-3, 69-70, 97, 126
geophysical exploration, 30, 31, 47, 99, 116
Geophysical Services, Incorporated, 106
getulista, 111
O Globo, 4, 139-40
Godinho, Cláudio Carlos, 163
Góes, Ismar de, 94
Góes Monteiro, General Pedro Aurélio de, 41
Gonçalves, Bento, 125
Goulart, João: Vice-President, 110, 111, 112, 137, 152; President, 104, 137, 138, 143, 149, 151, 152, 153, 155, 156, 157, 161, 162, 163, 164, 170, 171, 177
government: federal, 9, 12, 14, 15, 18, 19, 21, 23, 26, 27, 28, 35, 36, 40, 42, 54, 57, 59, 69, 80, 81, 86, 87, 96, 98, 99, 105, 111, 121, 141, 151, 154, 155, 156, 157, 165, 169, 172, 179, 183, 188 (*see also individual presidents by name*); state, 8, 9, 14, 17, 19, 21, 24, 26, 30, 39, 81, 88, 89, 94, 142 (*see also individual states by name*); municipal, 39, 81, 86, 88, 89, 94, 142, 154
government enterprise, 13, 15, 17, 18, 19, 21-2, 26, 27-8, 33, 34-5, 36, 40, 49, 51, 54, 61, 64, 75, 144, 168, 185. *See also* Petrobrás, Volta Redonda, *and individual mining and petroleum agencies by name*
graben, 28, 29
Grand Legislative Commission, 21
Grant, John Cameron, 8
Great Britain, 7, 54, 56, 62, 72. *See also* England
Guanabara: state, 66, 137, 148, 153, 156, 162, 170; bay, 114, 116
Gudin, Eugênio, 104, 105, 107, 150
Guinle, Guilherme, 31, 36
Gulf Oil Company, 110, 114, 157

Hanson's Latin American Letter, 122
Higher Institute of Brazilian Studies (ISEB), 59, 125

Higher War College (ESG), 58, 76, 100
Holland, 61
Hoover, Herbert, Jr., 62-3
Horta Barbosa, General Júlio Caetano, 33-4, 35, 49, 52, 53, 76, 113; CNP president, 37, 38, 41, 42, 43-4, 47, 54; resigns, 45, 46; campaigns for state monopoly, 54, 55, 58, 61, 63, 64, 65, 66, 72, 73, 76, 150, 167
hydro-electric power, 13, 18. *See also* electricity
Hydrocarbon Research Inc., 71

I. B. Sabbá & Company Ltd. (later Companhia de Petróleo da Amazônia), 92, 157
igneous rock, 3, 10, 15
Ilheus, Bahia, 7
imperialism, 60, 107, 109, 110, 118, 127, 154, 167
imports, 91, 117, 123, 129, 172; of petroleum, 42, 55, 70, 77, 79-80, 91, 103, 128, 131, 139, 146, 150, 157, 160, 163, 173, 174, 178; special exchange rates, 151, 172
import-substitution, 110
industrialists, 98, 143-4
industrialization, 48, 58, 59, 60, 66, 73, 76, 82, 98, 123, 167, 168, 186
industry, 13, 33, 44, 77, 140
inflation, 49, 72, 91, 120, 129, 172, 179, 180
inquiry commission, 29, 30. *See also* Petroleum Inquiry Commission
inquiry committee, 163, 172, 173; Chamber of Deputies, to investigate "Link Report", 134, 136; Congressional, 145, 150, 163
Integralista, 105
intellectuals, 156
International Monetary Fund (IMF), 120, 125
international oil companies, 1, 14, 26, 36, 39, 51, 53, 54, 55, 56, 57, 60, 62, 64, 66, 77, 91, 95, 103, 111, 119, 123, 124, 146, 154, 155, 161, 167, 183. *See also* "trusts", *and individual oil companies by name*

Interventor, 20-1, 63, 99
inventor, 14
Investments Committee, 62
Ipojuca, 170
Iran, 56
Irati, Paraná, 144
Italy, 58, 76, 110, 115, 166
Itamarati (Brazil's foreign ministry), 109, 110, 120
Itaparica, Bahia, 42

Jacuipe, 142
Jaguaré, São Paulo, 39
Jaguaribe, Hélio, 117, 125
Japan, 72, 179
Jofilly, José, 137
Joppert, Professor Maurício, 90, 95
O Jornal, 4, 63, 115, 121
Jornal de Debates, 56, 57, 59, 64, 108
Jornal do Brasil, 4, 17, 26, 27, 133, 140, 147, 148, 149, 153, 155, 157, 158, 161, 162, 163, 172, 181
Justice Committee, 19

Kellogg, W., Company, 52
Kennedy, John F., 144
Korean War, 76
Kubitschek, Juscelino, 104, 108, 110, 111, 112, 113, 114, 117, 119, 120, 127, 133, 141, 143, 151, 152
Kuwait, 157

labour, 110, 134, 137, 139, 140, 143, 146-9, 152, 153-4, 155, 156, 157, 161, 162, 170, 171. *See also* PTB, syndicate, *trabalhismo*, union
Labour, Minister of, 157
Lacerda, Carlos, 66, 84, 101, 111, 141, 153-4, 162
Lacerda, Hélio, 64
Lage, Henrique, 24
landowner (landholder, planter), 7, 8, 9, 10, 13, 14, 20, 21, 42, 46
Latin America, 52, 60, 62, 115, 167
lava, basaltic, 3
Law 2004, 96, 97, 99, 102, 105, 106, 109, 111, 114, 115, 119, 128, 142, 165, 174
Leal, General Estillac, 76, 85, 161
Lebanon, 122

"left", 140, 147, 162
legislation, petroleum, 36, 42, 43, 51, 52, 61-2, 80, 96, 109, 118. *See also* decree, Decree-Laws, Law 2004, Mining Code, National Petroleum Council, Petrobrás
legislative proposals (petroleum), 12, 17, 18, 19, 23, 46, 48, 53, 55-6, 61-74, 77, 78, 79, 82, 88, 94, 107, 108, 125. *See also* National Petroleum Institute, Petrobrás, Petroleum Statute
Leitão de Carvalho, General Estévão, 64-5
Leite, Solidónio, 17
Levy, Colonel (later General) Arthur, 52, 100, 103, 104, 147
liberalism, 4, 49, 52, 63, 66, 71, 72, 73, 74, 84, 85, 87, 88, 91, 93, 94, 96, 100, 104, 105, 107, 109, 113, 119, 120, 124, 125, 139, 151, 164, 169, 186; business, 90, 93-4, 112, 152. *See also* private enterprise
liberalization, 53, 55, 56, 58, 60, 95, 98, 125, 169
Lima, Hermes, 65-6
Lima, Mário, 151, 171
Lima e Silva, Professor Ruy de, 52
Lima Rocha, Heitor, 138, 139
Link, Theodore, 126
Link, Walter K., 125, 129, 135, 136, 137, 159, 160, 176, 181, 184; forms DEPEX, 100, 103; report to Sardenberg, 126; reputation, 126, 127, 128, 143, 144
"Link Report", 128-9, 133, 136, 139, 143, 158, 164, 165, 175, 183, 184
"Linkism", 128, 137, 143
Lins, Etelvino, 105
loans, 51, 144
Lobato, Bahia, 22, 29, 30, 31, 36, 42, 49, 107; Lobato-Joanes oilfield, 47; discovery of oil, 37, 100
Lobo Carneiro, Fernando Luiz, 56, 57, 59, 82, 83, 85, 86, 88, 90, 135
Loide Brasileiro, 154
Lott, General (later Marshal) Henrique Teixeira, 111-12, 117, 133, 141, 150
LPG (liquid petroleum gas), 144-5, 172

lubricating oils, 33, 91
Ludolf, Mário Leão, 140

Macedo Soares, José Carlos de, 119
Macedo Soares, José Eugênio de, 181
Maceió, Alagoas, 113
MacNaughton, Lewis W., 47, 68
Madeira River, 106
Mader, Othon, 93, 94
Magalhães, Juracy, 63, 64, 99-100, 103, 104, 136, 139
Maia, Alfredo, 9
Malamphy, Mark, 22, 23, 24, 30
Manaus, Amazônas, 92, 99, 106, 114, 128, 157
Mangabeira, Francisco, 153, 171; Petrobrás president, 139, 140, 143, 145, 146, 150, 151, 152; demands for dismissal as president, 146-9
Manguinhos, Federal District (later Guanabara), 51, 71, 114, 128, 157. *See also* Refinaria de Petróleo do Distrito Federal
manifesto, 28, 146-7
Maranhão: sedimentary basin, 3, 159; state, 8, 99, 171, 176
Maraú River, 7
Marighela, Carlos, 55
marine fossils (sediments), 2, 8
Marinho, Josafá, 134
Martins, Pinto, 15
Mascarenhas, Colonel Anderson, 140
Matarazzo, 50
Mataripe, Bahia, 51, 52, 70, 79, 91, 99, 101, 123, 131, 135, 142, 152, 158, 171, 178
Mattei, Enrico, 166
Mazzilli, Ranieri, 170
media, 4-5, 26, 28-30, 59, 63-4, 66, 84, 86, 90, 105, 109, 110, 118, 119, 122, 124, 126, 127, 129, 133, 143, 146, 153, 155, 157, 158, 159, 162, 188. *See also individual newspapers by name*
Melo Silva, Drault Ernani de, 51, 71
Mexico, 44, 90, 103; Constitution, 12; *1938* oil expropriation, 35, 54
Middle East, 61, 65, 122

Middle North, 3
military, 18, 20, 27, 33, 38, 52, 55, 57, 58, 64, 155; attitudes toward political activity, 75-6, 86, 141-2; nationalistic wing, 76, 101, 112; moderates, 85-6; and oil, 100, 105, 107, 109, 111, 121, 124, 141-2, 186. *See also* army, Clube Militar
military police, 172
Minas Gerais, 7, 9, 20, 66, 110, 131, 133, 178
Mineração e Metalurgia, 65
Mines and Energy, Minister of, 137, 138, 143, 144, 148, 179; Ministry of, 138-9, 177, 181
mining, 9, 14, 21, 23, 27, 33, 46; legislation, 7, 8, 13, 14, 21
mining code, 9, 21, 23, 27, 40, 41, 43-4, 46
mining school, 7
Miranda, Celso da Rocha, 145
Miranga, Bahia, 176
mixed company, 51, 54, 55, 56, 59, 61, 79, 84, 85, 105
modernization, 60, 78, 183
Monteiro Lobato, José Bento, 24, 25, 26, 27, 29, 30, 32, 33, 35, 37, 56
motor vehicles, 10, 43, 45, 48, 77, 80, 81, 83, 98, 129, 188
Moura, Pedro de, 25
Mourão Filho, General Olímpio, 170
Mundogás, 145
Murray-Simonsen, 36

naphtha, 175
National Bank of Economic Development (BNDE), 119-20, 125
National Directory of Mineral Production (DNPM), 23, 25, 26, 27, 28, 29, 30, 31, 32, 34, 36, 38
National Institute of Technology, 30
National Petroleum Council (CNP), 35, 37, 38, 39, 40, 42, 45, 46, 47, 48, 50, 52, 54, 56, 60, 69, 72, 78, 81, 87, 91, 93, 95, 98, 99, 100, 102, 103, 106, 108, 114, 115, 116, 117, 118, 119, 120, 127, 128, 129, 130, 131, 133, 134, 140, 146, 150-1, 163, 171, 178, 180, 186

National Petroleum Institute, 55-6
National Security Committee, 94
National Security Council, 62, 71, 109, 171, 174
National Student Union (UNE), 57, 88, 122, 137
nationalism, 1, 4, 19, 28, 34, 36, 39, 44, 50, 52, 53, 54, 60, 61, 66, 71, 73, 74, 75, 76, 78, 79, 80, 82, 88-9, 91, 92, 93, 95, 96, 98, 100, 104, 105, 107, 110, 111, 112, 113, 114, 115, 118, 119, 120, 122, 123, 128, 135, 137, 138, 139, 140, 142, 143, 146, 150, 151, 153, 155, 166, 168, 169, 183, 184, 185, 186. *See also* chauvinism, developmentalist nationalism, nativism, radical nationalism, xenophobia
Nationalist Parliamentary Front (FPN), 111, 123, 125, 128, 135, 137, 138, 152
nationalization, 35, 36, 54, 61, 62, 110, 111. *See also* distribution *and* refineries, private
nativism, 10, 12, 15, 59, 185, 186. *See also* chauvinism
natural gas, 42, 99, 159, 160, 174, 175, 176
natural resources, 21, 22, 58, 59, 74. *See also* water
Navy, 89, 109, 146; Ministry of, 154
neocolonialism, 167
Nery, Adalgisa, 139, 140, 141, 143, 144, 146
Neves, Brigadier Ari, 171
New York Times, 4, 91, 96, 114, 119, 144, 172
Northeast Coastal, 3
Nova Olinda, Amazônas, 99, 106, 107, 126, 142
NOVACAP, 154
Nunes, Colonel Janari, 108, 109, 112, 113, 115-16, 117, 118, 133

Obino, General Salvador César, 64
Odell, Peter, 60
officer corps, 76, 77, 85, 112, 186. *See also* army, military
Oil and Gas Journal, 15, 96, 126

Index 285

oil reserves, Brazil, myths about, 1-2, 4, 19, 31, 64, 69, 71, 77, 81, 90-1, 93, 121, 127, 131-2, 136, 164, 165, 167, 177, 182, 183-4, 185, 186, 187
oilfield, 18, 35, 53, 55, 61, 85, 91, 98, 99, 105, 121, 127, 136, 148, 159, 160, 175, 176, 177, 183, 188
Oliveira, Avelino Ignacio de, 36, 52
Oliveira, Euzebio Paulo de, 10, 12, 16, 17, 18-19, 22, 24, 25
Oliveira, Rafael Corrêa de, 56-7, 63
Oppenheim, Victor, 23, 24, 25, 29, 30, 31, 128-9
Otoni, Dr. Décio Savério, 106

Padhilha, Raimundo D., 67-9, 70, 90, 97
Paiva Teixeira, Glycon de, 23, 31, 34-5, 36, 49, 50, 99, 106, 134, 163
Paleozoic basins, 126
Pará, 99
paraffin, 47, 51, 52, 69, 135, 188
Paraguay, 55, 145
Paraná: oil province, 3; sedimentary basin, 10, 16, 17; state, 11, 37, 93, 99, 144
Passos, Gabriel de Rezende, 128, 134; divulges "Link Report", 126; Minister of Mines and Energy, 137, 138, 140, 143, 148
patriotism, 12, 32, 83, 105, 137, 150, 153, 156
Paula, R. Descartes de Garcia, 65
Paulinia, São Paulo, 178
Paz Estenssoro, Víctor, 107, 108
Pena, Affonso, 10
Pereira da Silva, 66
Pernambuco, 94
Persian Gulf, 145
Peru, 16, 61, 92, 114, 151
Pessoa, General José, 65
Petrobrás, 79-189 *passim*; bill creating (Projects 1516/51, 1517/51), 79-81; debate in Chamber of Deputies, 82-5, 87-90, 125; debate in Senate, 92-5; Deputies debate Senate amendments, 95; revenue, initial sources, 80-1, 99; revenue, ongoing, 81-2, 96-7, 128; share distribution, 80-1; passage of Law 2004, 95-6; management, 99-100; commences operations, 98-9, 102; organization, 103, 104, 117; focus (symbol) of nationalism, 101, 108, 111-13, 125; contracts with foreign concerns, 103; and Bolivia, 109-10; reaction to "Link Report", 126-8; *1959* financial study, 129-30; distribution of derivatives, 139, 146, 150-1, 154; monopoly over crude-oil imports, 131, 139, 146, 150, 157-8; *1969* financial analysis, 179-81; begins foreign exploration, 182-3, 184, 187; *1974* discoveries, 188-9
Petrobrás (magazine), 173
Petrobrás Internacional S.A. (Braspetro), 182
Petrobrás Química (Petroquisa), 175
petrochemicals, 102-3, 108, 114, 174, 175, 181
"O petróleo é nosso", 58, 59, 61, 65, 66, 72, 73, 75, 77, 84-5, 94, 121, 125, 147, 169-70, 181
petroleum: authorization for exploitation, 12, 23, 27, 48, 81; consumption, 47, 67, 70, 71, 77, 79, 80, 92, 97, 102, 103, 113, 118, 119, 124, 134, 135, 173, 181, 182, 188, 189; derivatives, *see* distribution of petroleum derivatives, tax; exploitation, *see* concession for petroleum exploitation; legislation, *see* legislation, petroleum *and* legislative proposals; quality, 47, 51, 175. *See also* production of petroleum
Petroleum Centre, 57, 63, 65, 72, 73, 76, 78, 82, 83, 86, 87, 95, 125
Petroleum Data Book, 67
petroleum industry, costs, 61, 67-9, 70, 77, 91, 97, 129-30, 141, 144, 160, 174
Petroleum Inquiry Commission (1936-37), 31, 32, 52
Petroleum Intelligence Weekly, 158

petroleum products, sales, 56, 83, 178, 179, 180
petroleum reserves, 53, 54, 55, 57, 61, 64, 65, 70, 74, 78, 85, 86, 91, 93, 112, 125, 131, 134, 164, 176, 177, 181, 182, 183; surveys, 125-6, 135, 168-70
Petroleum Scandal, The, 32
Petroleum Statute, 61-74 *passim*, 76, 77, 79, 81, 82, 84, 86, 87, 94, 97, 99, 185, 186
Petronal-Petróleo S.A., 145
Pimenta, Matos, 64
Pinheiro Neto, João, 140, 153
Pinto, Bilac, 84
pipeline, 69, 109, 116, 119
Pires, Waldir, 156
Pius XI (Pope), 64
political rights, suspension of, 171
Politics of Brazilian Nationalism, The, 38
pollution, 188-9
Polytechnical School of Bahia, 91
Pompeu, Plínio, 92, 94
Pôrto Alegre, Rio Grande do Sul, 153-4, 164
Portugal, 7
PR (Republican Party), 125
Prestes, Júlio, 16
Prestes, Luís Carlos, 121
price, 39, 61, 108, 116, 117, 129-30, 145, 172, 173, 174, 180, 189
private enterprise (companies), 11, 16, 18, 19, 22, 23, 24, 25, 28, 33, 34, 36, 37, 39, 40, 42, 49, 50, 53, 54, 60, 61, 64, 79, 82, 85, 87, 90, 92, 102, 109, 110, 118-21, 124, 125, 143, 144, 156, 160, 174, 175, 178, 183, 184
production of petroleum, 45, 47, 48, 52, 55, 67, 68, 69, 70, 72, 79, 91, 98, 99, 102, 106, 113, 115, 116, 118, 122, 124, 130, 131, 135, 138, 139, 141, 142, 143, 144, 145, 149, 153, 158, 159, 160, 165, 170, 174, 175, 176, 177, 179, 181, 182, 184, 188, 189
Provisional Government, 20, 21
PSB (Brazilian Socialist Party), 56, 137, 148, 150
PSD (Social Democratic Party), 85, 92, 94, 95, 105, 133, 137, 153
PSP (Social Progressive Party), 83, 85
PTB (Brazilian Labour Party), 73, 82, 85, 87, 88, 92, 95, 110, 133, 137, 152, 154. *See also* labour, *trabalhismo*
public opinion, 1-2, 4, 57, 62, 75, 77, 83, 84, 86, 90, 98, 108, 122, 124-5, 157, 186
public relations (Petrobrás), 115, 136, 142-3, 166, 170, 171, 172, 173, 177, 182
public utility, 35
purge, 146, 171

Quadragesimo Anno, 64
Quadros, Jânio, 104, 133, 134, 136, 137, 139, 152
Queirós, General Ademar de, 176; Petrobrás president, 170, 172, 173

radical nationalism, 4, 59, 64, 65, 72, 73, 81, 82, 83, 84, 85, 88, 89, 96, 104, 109, 111, 112, 119, 120, 121, 122, 125, 127, 128, 129, 134, 135, 137, 140, 141, 143, 145, 148, 151, 153, 155, 159, 164, 166, 170, 171, 185-6
railroads, 51, 77, 109, 118, 119
rationing, 43, 45, 49
Recôncavo, Bahia, 3, 31, 36, 37, 42, 45, 47, 48, 51, 67, 91, 92, 98, 99, 106, 113, 125, 126, 129, 138, 143, 148, 149, 154, 176
Refinaria de Petróleo do Distrito Federal, 71, 114. *See also* Manguinhos
Refinaria e Exploração de Petróleo União, 51, 71, 114, 119, 121, 141, 155. *See also* Capuava
Refinaria Nacional de Petróleo S.A., 52
refineries, government, 38-9, 48, 51, 52, 56, 70, 79, 91, 98
refineries, Petrobrás, 100, 102, 114-15, 116-17, 123, 129, 130, 131, 133, 135, 142, 147, 153, 156, 173, 177-8; adaptation to process national crude, 135. *See also individual refineries by name or location*

refineries, private, 39-40, 41, 48, 50, 51, 70, 83, 90, 94, 95, 97, 109, 113-14, 128, 129, 131, 135, 141, 173, 177; proposals to build, 48, 50, 51, 79, 82; nationalization campaign, 128, 134, 146, 148-9, 150, 152, 154-5, 156-7, 163, 164, 171, 173. See also individual refineries by name or location
refining, 33, 35, 37, 39, 54, 55, 61, 67, 68, 71, 72, 81, 84, 86, 90, 103, 113-15, 128, 181, 182, 184; capacity, 48, 50, 51, 79, 80, 91, 92, 94, 95, 98, 99, 102, 114, 124, 128, 131, 133, 134, 135, 177-8
reforms, 156, 163
Reis, Hugo Régis dos, 152, 158, 159, 161-2, 163, 171
Republic, First (Old), 8, 19, 20, 21
revenue (income, funds), 56, 69, 77, 78, 80, 81, 82, 84, 96-7, 99, 102, 116, 129-30, 150, 165, 166, 179
Revista Brasileira de Geografia, 36
Revista do Clube de Engenharia, 65, 90
Revista do Clube Militar, 64-5, 76, 85-6, 141
Revolution of 1930, 19, 20, 53
Riacho Doce, Alagoas, 11, 23, 24, 25, 30
Rio de Janeiro: city (Federal District, later Guanabara), 4, 15, 36, 48, 51, 58, 63, 64, 71, 79, 88, 90, 104, 113-14, 119, 122, 125-6, 133, 136, 137, 139, 146, 157, 163, 170; state, 39, 114, 137, 156
Rio Grande do Norte, 16
Rio Grande do Sul, 10, 20, 50, 84, 114, 131, 153-4, 170, 177, 178
Roboré, Bolivia, 118; Accord, 118-20, 153
Rocha, Euzébio, 66, 82, 83, 85, 87, 88-9, 94, 95
Rockefeller "group", 145
Rodrigues Alves, Francisco de Paula, 9
Romero, F. B., 24-5, 26
Roxo, Mathias G., 31
Royal Dutch–Shell (Anglo-Mexican), 21, 32, 157-8, 178
royalties, 163

Rumania, 159
"Russian Report", 158-60

sabotage, 26, 31-2, 127, 136, 141, 159, 171
Salgado, Plínio, 105
SALTE plan, 70-1, 72, 79, 91, 99
Salvador, Bahia, 7, 22, 48, 51, 136
Santa Catarina, 8, 10
Santa Cruz de la Sierra, Bolivia, 118
Santos, Max Costa, 150
Santos, São Paulo, 34, 109, 114
Santos-São Paulo pipeline, 103
São Borja, Rio Grande do Sul, 73
São Luís, Maranhão, 176
São Mateus, Paraná, 144, 174, 177
São Paulo: city, 20, 36, 48, 50, 51, 71, 90, 113-14, 124, 151, 156; state, 8, 9, 11, 15, 16, 17, 20, 25, 26, 39, 66, 67, 79, 88, 98, 99, 133, 141, 178
São Pedro, São Paulo, 25, 29
Sardenberg, Colonel (later General) Idálio, 134, 135; Petrobrás president, 117, 123, 126, 128, 133, 147
Sargent, Thomas Denny, 7
Saudi Arabia, 57, 114, 158
schists (shales), 8, 31, 174, 177, 181, 189; *1962* proposals to develop, 143-4
Second World War, 4, 42, 47, 49, 51, 54, 58, 62, 73, 76, 79, 93, 96, 168, 185, 186
secondary recovery, 159, 160, 176-7
security, national, 18, 19, 27, 38, 44, 48, 58, 65, 152, 182, 183, 185, 186
sedimentary basins, 2-4, 14, 15, 26, 31, 103, 106, 126, 136, 159, 184. See also Paleozoic basins
seepage, oil, 7-8, 22, 31
self-sufficiency, 67, 68, 69, 70, 78, 97, 100, 101, 103, 112, 113, 118, 124, 127, 133, 135, 143, 158, 166, 170, 177-8, 180, 181, 184, 188, 189
O Semanário, 59, 108, 111, 112, 128, 137, 159
Senate, 66, 82, 88, 90, 92-5, 98, 105, 107, 164
Sergipe, 19, 40, 42, 99, 126, 171, 175, 176, 177

Service for the Stimulation of Mineral Production (SFPM), 23, 24, 28, 29, 30, 36
shares (stock), 46, 52, 80, 81, 82, 83, 88, 94, 95, 96, 129, 156, 165, 175, 180
Shell-Mex Brasil, 61
Silva, General Albino, 151-2, 153, 155, 157, 158, 160-2, 163
Silva, Amauri, 157
Silveira, Joel, 108
Silveira, Paulo, 138, 143
Simões Lopes, Ildefonso, 14, 15, 18
Simonsen, Mário Wallace, 145
Sinclair Refining Co., 157
Skidmore, Thomas E., 59, 155
slavery, abolition of, 83, 108
Soares Sampaio, Alberto, 51, 71
Sobrinho, Barbosa Lima, 131
socialism, 65, 139. *See also* PSB
Socony-Vacuum Oil Company Inc., 79
Solicitor General of Brazil, 156
South America, 182
Southwestern Engineering Company, 114
Souza, General Antônio José Alves de, 52
sovereignty, 17, 18, 44, 58, 66, 74, 90, 92, 109, 121
Soviet Union (Russia), 56, 62, 131, 144, 157-8, 174; petroleum technicians' assessment of Brazilian oil industry, 158-60
stabilization program, 120. *See also* inflation, International Monetary Fund
Standard Oil, 57, 64, 82, 83, 127, 136, 157, 161; of Brazil, 40, 178; proposals to develop Brazilian oil, 40-1, 43, 44, 49, 61; of California, 51, 114-15; of New Jersey, 17, 21, 32, 39, 40, 51, 78, 100, 106, 114, 123, 126, 145
state control, 21, 28, 35, 36, 49, 52, 53, 58, 59, 61, 62, 64, 71, 77, 81, 82, 100, 112, 167, 168, 185, 186. *See also* developmentalist nationalism, liberalism, nationalism, radical nationalism

state monopoly, 38, 53, 54, 55, 56, 57, 58, 63, 65, 66, 67, 83, 84, 85, 86, 87, 88, 89, 90, 93, 94, 95, 98, 100, 104, 105, 106, 108, 119, 121, 122, 123, 124, 125, 137, 146, 147, 150, 154, 155, 163, 167, 183, 184, 185, 186
steel, 39, 69, 78, 167-8
strike, 138, 141, 146-7, 149, 151, 155, 156-7, 170
students, 58, 59, 86, 88, 89, 111, 113, 125, 137, 154, 156, 169, 172. *See also* National Student Union
Subandean Zone, 34
subsoil, 8, 9, 13, 14, 18, 27, 36, 46, 48, 69, 77, 95, 140, 177
subversion, 172-3
Suez, 122
sugar, 13
sulphur, 9, 188-9
Sun Oil Company, 158, 160
Superintendency of Money and Credit, 160
survey, 9, 15, 16, 18, 26, 104
Sweden, 72
syndicate, 146, 147, 148, 151, 161, 162, 172. *See also* labour, union

Tabuleiro dos Martins, Alagoas, 143
tankers, 42, 43, 44, 56, 70, 71, 72, 99, 103, 116, 117, 122, 123, 130, 144, 170, 179; Petrobrás orders from Brazilian yards, 142, 179. *See also* FRONAPE
tanks, storage, 47, 69, 116, 149, 175
tariff, 33, 80, 118, 129-30, 165
Távora, Juarez, 27, 83-4, 100, 105-6, 107; Minister of Agriculture, 23, 29, 30, 53; supports liberalization of oil laws, 53, 54, 55, 58, 60, 61, 62, 63, 64, 65, 66, 86
tax, 39, 77, 78, 110, 165; single (sole, unified) federal tax on petroleum derivatives, 38, 80, 129-30, 180
technicians, 13, 18, 23, 30, 47, 54, 68, 79, 158-60, 162, 178; of Petrobrás, 133, 136, 146-7, 148, 157, 181
technocrats, 78, 89
tectonics, 3, 31
Tendler, Judith, 168-9

Index 289

tenente, 27, 53, 63, 99
Teresina, Ceará, 86
Texas Company, 21, 114, 158, 178
topping, 47
trabalhismo, 74. *See also getulista*, PTB, Vargas, Getúlio
Transport, Minister of, 162
transport, 18, 41, 44, 53, 54, 67, 69, 72, 81, 84, 103, 109, 130, 131, 160, 172, 181, 184
"trusts", 26, 29, 30, 32, 33, 46, 51, 55, 56, 57, 60, 61, 63, 65, 66, 72, 73, 79, 82, 83, 86, 89, 91, 93, 95, 96, 107, 108, 110, 112, 114, 115, 117, 118, 119, 122, 123, 124, 127, 129, 132, 135, 141, 144, 145, 152, 160, 163, 164, 165, 166, 167, 168, 169, 177, 183, 184. *See also* international oil companies
Tucano, Bahia, 3, 126, 143, 159

UDN (National Democratic Union), 66, 75, 87, 90, 92, 93, 95, 100, 105, 111, 125, 133, 169; policy change on Petrobrás, 83-5
Última Hora, 4, 59, 112, 117, 119, 122, 124, 131, 134, 138, 139, 140, 143, 150, 153, 156, 157, 158, 161
union (oil), 134, 138, 146, 151, 162, 163, 170, 171. *See also* labour, syndicate
United Geophysical Company, 47
United Nations, 58
United States of America, 13, 17, 18, 38, 42, 44, 47, 51, 54, 56, 57, 58, 62, 63, 64, 66, 67, 68, 69, 70, 71, 76, 78, 86, 96, 115, 119, 120, 122, 128, 144, 149, 155, 159, 188
University Reform Movement, 58
University of Brazil, 79
Uruguay, 38, 39, 40

Vargas, Getúlio, 24, 25, 27, 30, 31, 38, 40, 42, 43-4, 45, 46, 49, 66, 73, 77, 79, 86, 103, 104, 112, 137, 141, 152, 169, 186; Revolution of *1930*, 20, 21; Estado Nôvo, 33, 38, 58, 60, 66, 75, 76, 169; as oil nationalist, 34, 35, 44, 100; *1950* election, 74, 75, 76, 78, 110; and Petrobrás, 78, 79-82, 84-5, 87-9, 92, 95, 96, 98, 99, 100; suicide (1954), 44, 84, 101, 104, 107, 110, 113, 143, 161, 169; legacy for oil industry, 107-8, 113. *See also getulista*, National Petroleum Council, Petrobrás, PTB, *trabalhismo*
Vargas, Lutero, 137
Venezuela, 14, 41, 54-5, 61, 102, 114, 123, 157-8, 160
vertical integration, 77, 103, 124, 154, 170, 184
Viana, General Segadas, 141, 147
Virginia Geological Service, 10
Volta Redonda, Rio de Janeiro, 44, 167-8

Walter J. Levy, Inc., 129, 165, 179
War: Minister of, 44, 85, 111, 117, 141, 147, 170, 176; Ministry of, 45, 100
Washburne, Dr. Chester W., 15, 16, 25
Washington, D.C., 99
water, 9, 21, 142, 170
Western Hemisphere, 49, 57
wheat, 151
White, Israel Charles, 9-10, 16
Wilson, Edward Pellew, 7-8
Wilson, Woodrow, 55
Wirth, John D., 38
World Petroleum, 40, 96
World Petroleum Congress, Fifth, 126
World Petroleum Report, 135

xenophobia, 19, 61, 63, 183, 185. *See also* chauvinism, nativism

Yacimientos Petrolíferos Fiscales (YPF), 35, 38, 145
Yacimientos Petrolíferos Fiscales Bolivianos (YPFB), 107, 118
Ypiranga (Companhia de Petróleos Rio Grande), 114, 157, 178
Yugoslavia, 179

SAM/PSE E/I 123 PAB
            OO
         Smith